PHANTASMATIC SHAKESPEARE

PHANTASMATIC SHAKESPEARE

IMAGINATION IN THE AGE OF EARLY MODERN SCIENCE

SUPARNA ROYCHOUDHURY

CORNELL UNIVERSITY PRESS

Ithaca and London

First published 2018 by Cornell University Press

Printed in the United States of America

Library of Congress Cataloging-in-Publication Data

Names: Roychoudhury, Suparna, author.
Title: Phantasmatic Shakespeare : imagination in the age
    of early modern science / Suparna Roychoudhury.
Description: Ithaca : Cornell University Press, 2018. |
    Includes bibliographical references and index.
Identifiers: LCCN 2018016351 (print) | LCCN 2018018663
    (ebook) | ISBN 9781501726569 (pdf) |
    ISBN 9781501726576 (ret) | ISBN 9781501726552 |
    ISBN 9781501726552 (cloth ; alk. paper)
Subjects: LCSH: Shakespeare, William, 1564–1616—
    Criticism and interpretation. | Literature and
    science—England—History—16th century. | Literature
    and science—England—History—17th century. |
    Science in literature. | Imagination in literature. |
    Psychology in literature. | Cognition in literature.
Classification: LCC PR3047 (ebook) | LCC PR3047 .R69
    2018 (print) | DDC 822.3/3—dc23
LC record available at https://lccn.loc.gov/2018016351

For my parents

CONTENTS

Illustrations

ACKNOWLEDGMENTS

Many people helped in the imagining of this book. For its phantasmatic beginnings, I thank Stephen Greenblatt, whose provocative responses shaped its initial form, as did the searching and inspiring critiques of Marjorie Garber and Elaine Scarry. I have learned from many readers and auditors at the Shakespeare Association of America, the Renaissance Society of America, the Modern Language Association, and the Mahindra Humanities Center at Harvard University; their reactions and questions have improved the book in innumerable ways. Among them are Reid Barbour, Dympna Callaghan, Darryl Chalk, Mary Crane, Ananya Dutta Gupta, Mary Floyd-Wilson, Simon Goldhill, Rachel Holmes, Sachiko Kusukawa, Alexander Marr, Alan Richardson, Debapriya Sarkar, Laurie Shannon, Stephanie Shirilan, Virginia Strain, Richard Strier, Elizabeth Swann, Rebecca Totaro, Douglas Trevor, Lyn Tribble, Henry Turner, Travis Williams, and Jessica Wolfe. Many more crucial contributions were made informally, by all those friends and colleagues who offered to think, read, and talk with me. They include Chris Barrett, Liza Blake, Kim Coles, Jane Degenhardt, Rachel Eisendrath, Susan Gaylard, Jamey Graham, Diana Henderson, Jean Howard, James Knapp, Erika Lin, Amy Rodgers, Yulia Ryzhik, Jonathan Sawday, Sarah Wall-Randell, Michael Witmore, and Adam Zucker.

I could not have written this book without the generous support of several institutions. I am obliged to the Newberry Library for a long-term fellowship funded by the National Endowment for the Humanities; to the University of Cambridge for a fellowship at the Centre for Research in the Arts, Social Sciences, and Humanities; and to the Folger Shakespeare Library. Earlier on I received grants from Harvard University; in the latter stages I was supported by awards from Mount Holyoke College. I cannot forget the many librarians and archivists who assisted me over the years, at the Newberry and the Folger as well as the British, Bodleian, Wellcome, and Huntington Libraries. Parts of this book originally appeared elsewhere: an earlier version of chapter 5 was published under its present title in *Modern Philology* 111, no. 2 (November 2013), 205–30, and an abbreviated version of chapter 1 appeared as "Anatomies of Imagination in Shakespeare's Sonnets" in *SEL Studies in English Literature 1500–1900* 54, no. 1 (Winter 2014), 105–24. For these publication opportunities I acknowledge in particular Richard Strier and Logan Browning, respectively. At Cornell University Press, I have benefited greatly from the incisive remarks of Jenny Mann, who read the manuscript, and the wise interventions of my anonymous reviewer. I am deeply grateful to my editor Mahinder Kingra for his intuitions and guidance, and thank all the members of my skilled editorial and production team, including Bethany Wasik, Sara Ferguson, Michelle Witkowski, and Rachel Lyon.

My thanks, finally, to the members of the Harvard Renaissance Colloquium, the earliest friends of this project. I salute my sister Supriya and extended family, my English department colleagues at Mount Holyoke, and my students, all of whom sustain my work in indirect but meaningful ways. And I acknowledge the special debt I owe the late Iain Wright, who persuaded me to embark on the academic adventure that transformed my life. For their encouragements and love, I dedicate this book to my parents, Partha Sarathi Roychoudhury and Sanjukta Roychoudhury.

PHANTASMATIC SHAKESPEARE

Introduction

Theseus, *Phantasia*, and the Scientific Renaissance

Representations of the imagining mind have a central place in Shakespeare's art. Instances of perceiving and conceiving pervade the plays and poems—we see Bottom struggling to recount a magical dream, Macbeth reaching for a dagger that is not there, Prospero humbling his enemies with wondrous illusions, the speaker of the sonnets recalling the face of his beloved. These and countless other moments naturally prompt the question of what Shakespeare knew about the cognitive process of mental representation. In a sense, the question is easy to answer. The prevailing model of psychology in late sixteenth-century Europe, we know, held that the mind consists of several powers or faculties. Of these, imagination is one; specifically, it is the faculty of the sensitive soul that creates and manipulates phantasms—intelligible mental forms. With the aid of phantasms, imagination, along with the other faculties of understanding and memory, carries out essential functions of mental cognition. As is evident, Shakespeare knew the principles of faculty psychology. Yet, as I mean to show in this book, his representations of imagination do not merely echo those principles. Rather, they foreground ambiguities in the early modern conception of imagination that arose around the same time as important shifts in early modern science—in such fields as anatomy, medicine, mathematics, and natural history. Shakespeare's portrayal of imagination draws on the uniquely idiosyncratic discourse of the image-making mind; those idiosyncrasies in turn connect with the work of contemporary natural philosophers and protoscientists such as Andreas Vesalius, Johannes Kepler, and Francis Bacon. The interdiscursive impact that science had on faculty psychology is registered in various ways throughout Shakespeare's corpus. Adapting the intellectual tensions of his time to literary form, Shakespeare brings the increasingly ambiguous epistemology of imagination firmly into the purview of art. Imagination in Shakespeare, then, is less the wellspring of artistic production than a set of problems that require representation for their expression. Situating Shakespeare's depictions of imagination amid the epistemic transitions of his time, in this book I contend that the power of those depictions draws from the newly destabilized and therefore dynamically generative idea of imagination that prevailed in Renaissance culture.

A reading of Shakespeare's best-known pronouncement on imagination, a brief exchange between Theseus and Hippolyta near the end of *A Midsummer Night's Dream*, will help introduce the overarching concerns of this study. The variegated and ambivalent characterization of imagination that we see here is deeply steeped in the rich psychological discourse of the late sixteenth century. Not only does Shakespeare allude to premodern intellectual traditions of imagination, of which Aristotelian faculty psychology will be our special focus; he also gestures toward the broader philosophical and protoscientific attitudes of the age. In a short space, Shakespeare broaches the particular facets of imagination that would occupy him throughout his career—and that I will explore in greater depth in the subsequent chapters of this book.

Pondering the bizarre report of the lovers' midsummer adventure, Hippolyta remarks, "'Tis strange my Theseus, that these lovers speak of." To which he replies,

> More strange than true. I never may believe
> These antique fables, nor these fairy toys.
> Lovers and madmen have such seething brains,
> Such shaping fantasies, that apprehend more
> Than cool reason ever comprehends.
> The lunatic, the lover, and the poet
> Are of imagination all compact.
> One sees more devils than vast hell can hold;
> That is the madman. The lover, all as frantic,
> Sees Helen's beauty in a brow of Egypt.
> The poet's eye, in a fine frenzy rolling,
> Doth glance from heaven to earth, from earth to heaven.

And as imagination bodies forth
The forms of things unknown, the poet's pen
Turns them to shapes and gives to airy nothing
A local habitation and a name.
Such tricks hath strong imagination
That if it would but apprehend some joy,
It comprehends some bringer of that joy;
Or in the night, imagining some fear,
How easy is a bush supposed a bear![1]

Theseus groups "the lunatic, the lover, and the poet" in a trinity of imagination's hapless victims. Of lovers we have seen much already in this play. We
have also witnessed some measure of lunacy in the delusions produced by
Puck's flower. But the invocation of the "poet's pen" is less expected and therefore striking; maybe because of its sudden intrusion, the speech continues to
be read as an articulation of Shakespeare's beliefs about poesy, as praise for
what C. L. Barber terms "the power of the mind to create." Such a reading is
somewhat undercut, however, by the ungenerous elements of Theseus's characterization. He conceives of imagination as next door to insanity and indig-
nity; as Marjorie Garber says, he "discounts and criticizes art."[2] Shakespeare
also signals that we should not take this speech at face value. The speaker's
obliviousness, for instance, alerts us to the presence of a subtext: Theseus,
slayer of the Minotaur in Greek mythology, does not know that he himself
is an "antique fable"; or that he, an impatient bridegroom, is a "lover" too; or
that there really are "fairy toys" in the woods outside Athens.

Theseus's "poet" is in fact a belated addition to what was initially a duo—
"lovers and madmen." There are many more dyads and triads, more structures
and hierarchies, that Theseus tries out on the fly. There is the cosmological
triple of hell, heaven, and earth, as well as the creaturely one of spirits, humans,
and beasts. "Helen's beauty" and the "brow of Egypt" are yoked in antithesis,
while "bush" and "bear" alliteratively conjoin the realms of flora and fauna. Implicitly, imagination is stitched into all the universe, involved in our understanding of nature as much as art. Still, the various categories are not presented
in an orderly way; their tumbling succession effects not so much a grand world
picture as a kind of quivering overdetermination. The declarative tone is

1. William Shakespeare, *A Midsummer Night's Dream*, 5.1.2–22. Shakespeare references are from
The Norton Shakespeare, ed. Stephen Greenblatt et al. (New York: W. W. Norton, 2016), hereafter cited
parenthetically.
2. C. L. Barber, *Shakespeare's Festive Comedy* (Princeton, NJ: Princeton University Press, 1972),
142; Marjorie Garber, *Shakespeare After All* (New York: Pantheon, 2004), 222.

undercut by a meandering quality: from the depths of "vast hell" we ascend to "heaven," then move "from heaven to earth." We travel from Greece to Egypt—or, if "Egypt" is glossed as "gypsy," into the unmarked territory of the nomad. Initially, it is said that "fantasies" apprehend and "reason" comprehends, but a few lines later "strong imagination" can do both: "if it would but apprehend some joy," it "comprehends some bringer of that joy." When it seems, with the repetition of "apprehend" and "comprehend," that we are drawing to a close, Theseus darts in another direction: "How easy is a bush supposed a bear!" Rhetorically, the speech is littered with cryptic adjective-noun pairs that tug us in different directions. "Strong imagination," Theseus says, consists in "antique fables" and "fairy toys." Imagination manifests in "seething brains" and "shaping fantasies." Its power exceeds "cool reason." It is an experience of "fine frenzy"; its provenance is "airy nothing." All at once, an imagination is a tall tale, a supernatural plaything, a disordered organ of the body, a formative principle, the antithesis of intellect, a flimsy substance. How are we to reconcile all these ideas?

Theseus's intuitions about imagination are mottled and diffuse, concerned not just with poetic expression but with a wide range of questions about knowledge and nature. If there is an intention to systematize and delineate, it has been disrupted by a proliferation of too many ordering principles. Theseus may be unconscious of the subtle tensions infusing his account; Shakespeare is not. When we return to this speech at the end of this chapter, I will suggest that compressed in it is a not insignificant quantity of premodern cognitive theory. There are indications as well of several areas in which imagination would be tested by new questions and problems emanating from new epistemes and epistemologies. What Theseus offers is not a unified theory of imagination. More accurately, it is a suspension of allusions, a pastiche conveying the instability of imagination as an idea. With this speech, Shakespeare does more than translate early modern psychological theory into poetry; he tacitly presents this act of translation as the very definition of poetry. Before we can unpack Theseus's various references—and parse Hippolyta's response—we need first to consider the contemporary discourse of imagination and the ways in which it was infiltrated by scientific and epistemological thinking.

Faculty Psychology and Early Modern Science

The rich premodern intellectual history of imagination has been traced by numerous scholars, including, in the earliest instances, Murray Wright Bundy and E. Ruth Harvey, and later by scholars such as Katharine Park and Stuart

Clark.[3] Broadly, the history may be said to fall into three streams: one originating with Plato, which emphasizes imagination's mystical and ethical implications; one following Aristotle, which is focused on epistemology and psychology, concerned with the empirical world of the senses; and a third, which is interested in imagination's aesthetic and rhetorical applications. This is a necessarily simplified view: in the postclassical era, these three traditions— ethical, epistemological, and aesthetic—are convolved with one another. Certainly, they are not clearly delineated as such in early modern culture. From the viewpoint of modernity, mysticism and aestheticism are imagination's perhaps more familiar characteristics, in part because of the influential post-Romantic association of imagination with inspiration and creativity—with the mysteries of genius and artistic excellence. My objective in this book, however, is to underline the importance that the epistemological tradition held for Shakespeare. Although this tradition focuses on imagination as a cognitive rather than a creative power, it had certain impact on Shakespeare's literary aesthetic.

Phantasia, the Greek name for imagination—derived from the verb *phantazein*, "to make apparent"—is scattered through the Platonic dialogues. In the *Timaeus* and *Phaedrus*, it is associated with illusion and the obfuscation of knowledge, on the one hand, and with divine inspiration and spiritual ascendance, on the other.[4] In Neoplatonist philosophy, imagination enables the mind to grasp the transcendent reality; at the same time, it is entangled with the carnal world of the senses. Plotinus resolves this difficulty by conceiving of a lower and a higher form of imagination: one "looks below to receive images of matter," while the other looks "above to receive images of thought" from higher planes.[5] In a not dissimilar way, the aesthetic utility of imagination

---

3. For overviews of premodern psychology, see Murray Wright Bundy, *The Theory of Imagination in Classical and Medieval Thought* (Urbana: University of Illinois Press, 1927); E. Ruth Harvey, *The Inward Wits: Psychological Theory in the Middle Ages and the Renaissance* (London: Warburg Institute, 1975); Eva T. H. Brann, *The World of the Imagination: Sum and Substance* (Savage, MD: Rowman and Littlefield, 1991); J. M. Cocking, *Imagination: A Study in the History of Ideas*, ed. Penelope Murray (London: Routledge, 1991); Simon Kemp, *Cognitive Psychology in the Middle Ages* (Westport, CT: Greenwood, 1996); Robert Pasnau, *Theories of Cognition in the Later Middle Ages* (Cambridge: Cambridge University Press, 1997); and Alexander M. Schlutz, *Mind's World: Imagination and Subjectivity from Descartes to Romanticism* (Seattle: University of Washington Press, 2009), 15–35. On the Renaissance reception of faculty psychology, see Katharine Park, "The Organic Soul," in *Cambridge History of Renaissance Philosophy*, ed. C. B. Schmitt et al. (Cambridge: Cambridge University Press, 1988), 464–84; and Stuart Clark, *Vanities of the Eye: Vision in Early Modern European Culture* (Oxford: Oxford University Press, 2007), esp. chaps. 1 and 2.

4. For a discussion of Plato's ideas about *phantasia*, see Bundy, *Theory of Imagination*, chap. 2, "Plato," 19–59.

5. Brann, *World of Imagination*, 49. See as well Plotinus, *Ennead, Volume IV*, trans. A. H. Armstrong, Loeb Classical Library 443 (Cambridge, MA: Harvard University Press, 1984), 4.3.30–31.

would also be cleaved in two. While Plato does not systematically connect *phantasia* with artistic production, he draws a distinction in the *Sophist* that would prove influential later on. There are two ways that an artificer uses imagination: to make likenesses that are faithful imitations of truth, or to fashion appearances that are distortive misrepresentations of truth.[6] This evaluative approach laid the grounds for future thinking about effective art. In early aesthetics, the careful use of imagination was tied to the creation of lifelikeness and sublimity. Philostratus praises the mind's power to imitate; Quintilian describes how the best orators rouse emotions in their audience with the help of vivid mental images; Longinus writes that the manipulation of imagery invests writing with grandeur and weight.[7]

The attitude of judgment carries through to Renaissance discussions of literary and aesthetic theory. For example, the question of whether poetic distinction lies in the imitations of the icastic imagination or in the fantastic illusions wrought by the fabulous imagination was much debated by sixteenth-century thinkers: the philosopher Jacopo Mazzoni avers that "phantasy is the true power over poetic fables," whereas Torquato Tasso says that "the best poetry imitates the things that are, that were, and that can be."[8] Among the English texts that reproduce this dichotomy of the mimetic versus the inventive imagination we would count Philip Sidney's *Defence of Poesy* and George Puttenham's *Art of English Poesy*. As scholars have pointed out, Renaissance poets strategically drew on the ancient philosophy of imagination in order to legitimize poesy as a worthwhile practice. Their task, which they carried out successfully, was to separate imagination's unflattering aspects from its more productive ones, demonstrating that, when properly disciplined, it could be "deliberate and purposeful, moral and rational," could persuade to good and reveal higher truths.[9] Over the course of the period, imagination and invention would gradually gain respectability in literary culture.

---

6. Plato, *Theaetetus. Sophist*, trans. Harold North Fowler, Loeb Classical Library 123 (Cambridge, MA: Harvard University Press, 1967), 235d–236c.

7. Philostratus, *Apollonius of Tyana, Volume I: Life of Apollonius of Tyana, Books 1–4*, ed. and trans. Christopher P. Jones, Loeb Classical Library 16 (Cambridge, MA: Harvard University Press, 2005), 2.22.3; Quintilian, *The Orator's Education, Volume III: Books 6–8*, trans. and ed. Donald A. Russell, Loeb Classical Library 126 (Cambridge, MA: Harvard University Press, 2001), 6.2.30–32; Longinus, *On the Sublime*, trans. W. Hamilton Fyfe, rev. Donald Russell in Aristotle, Longinus, Demetrius, *Poetics. Longinus: On the Sublime. Demetrius: On Style*, trans. Stephen Halliwell, W. Hamilton Fyfe, Doreen C. Innes, and W. Rhys Roberts, rev. Donald Russell, Loeb Classical Library 199 (Cambridge, MA: Harvard University Press, 1995), 15.1–2.

8. Quoted in Allan H. Gilbert, ed., *Literary Criticism: Plato to Dryden* (New York: American Book Company, 1940), 387, 475.

9. William Rossky, "Imagination in the English Renaissance: Psychology and Poetic," *Studies in the Renaissance 5* (1958): 65. See as well Idris Baker McElveen, *Shakespeare and Renaissance Concepts of the Imagination* (Ann Arbor: University Microfilms, 1979); and Peter Mack, "Early Modern Ideas of

Alongside these lines of thinking, however, lay an extensive epistemological discourse, a discourse not primarily interested in the moral or artistic merits of imagination—in prescribing how imagination ought to be used, ought to be valued—but rather more concerned with questions about function—what imagination does and how. These types of questions derive from Aristotelian philosophy. In his *De anima*, Aristotle used *phantasia* to name that power of the sensitive soul that forms the necessary bridge between sensation and judgment. The cognitive interface between world and mind, imagination is "the process by which we say that an image [*phantasma*] is presented to us." Another crucial idea is that the soul does all its thinking with the aid of *phantasmata*, or mental representations.[10] Over the course of the Middle Ages, Aristotle's suggestive remarks would be developed by others into a full-fledged cognitive theory, culminating in the system of faculty psychology that was inherited by the early moderns.

Islamic philosophers such as Avicenna (Ibn Sīnā) and Averroes (Ibn Rushd) expanded the Aristotelian system by delineating more clearly the different powers of the soul. Avicenna divided *phantasia* into three parts—the *sensus communis*, or common sense, which consolidates information gleaned from the physical senses; a retentive image-making power, which temporarily stores this information; and a compositive power, which manipulates phantasms in isolation of sense experience. The nomenclature grew nuanced: the Greek *phantasia* and the Latin *imaginatio*, which had once been equivalent, were differentiated; the latter represented the more mechanical operations of the image-making faculty, the former its free play.[11] Another new idea was that the various "inward" senses reside in different regions of the brain.[12] In the European Middle Ages, the differences between the faculties would be further crystallized and their individual functions pondered at length. Imagination grew in importance: Albertus Magnus, for instance, held imagination to be responsible for a wide spectrum of activities—retaining, creating, representing, even some forms of understanding.[13] Later, Thomas Aquinas would emphasize the Aristotelian axiom that the soul understands nothing without the aid of phantasms; only by engaging with the sensible world can the mind arrive at abstract truths.[14]

Imagination: The Rhetorical Tradition," in *Imagination in the Later Middle Ages and Early Modern Times*, ed. Lodi Nauta and Detlev Pätzold (Louvain, Belgium: Peeters, 2004), 59–76.

10. Aristotle, *On the Soul*, in *On the Soul. Parva Naturalia. On Breath*, trans. W. S. Hett, Loeb Classical Library 288 (Cambridge, MA: Harvard University Press, 1957), 427b15–18, 428a1–5.

11. Bundy, *Theory of Imagination*, 183.

12. Jon McGinnis, *Avicenna* (Oxford: Oxford University Press, 2010), 111–16.

13. See Bundy, *Theory of Imagination*, 190, 266; Brann, *World of Imagination*, 58–59.

14. Thomas Aquinas, *On Human Nature*, ed. Thomas S. Hibbs (Indianapolis, IN: Hackett, 1999), question 84, article 7, 151–52. See also Harvey, *Inward Wits*, 58.

In short, faculty psychology is a theory of cognition that supposes that all mental activity comprises the production and examination of mental representations. It understands the embodied mind to consist of several powers or faculties: powers that derive phantasms from sense impressions (common sense, fantasy, imagination); those that extract abstract ideas from such phantasms (reason, understanding); and that which retains them for subsequent use (memory). The cognitive function of imagination, therefore, is of central importance: sensing, conceiving, thinking, and remembering are contingent on its proper operation.

While the moral and aesthetic failings of imagination continued to be emphasized by such thinkers as Marsilio Ficino and Gianfrancesco Pico della Mirandola, it fell to faculty psychology to provide a systematic account of imagination's behavior. Still, it bears repeating that the ethical, epistemological, and aesthetic aspects of imagination are not segregated in Renaissance thought: if Aristotle's model supplies the explanatory framework, the explanations were infused with ideologies drawn from elsewhere. In focusing particularly on the cognitive mechanics of imagination, it is not my intention to suggest that Shakespeare was uninterested in evaluative or aesthetic considerations. My contention, rather, is that Shakespeare's representations of imagination are more anchored in epistemological problem solving than we have tended to assume, and that Shakespeare aestheticizes a way of thinking about imagination that was not originally conceived in aesthetic terms. Faculty psychology is also the stream of imaginative theory that would be most directly jolted by the novel epistemologies that emerged as part of the so-called scientific revolution. Reorienting our view of the Renaissance imagination so as to acknowledge more fully its nonartistic facets indicates that Shakespeare thought as much like a cognitive psychologist as he did a prophet or rhapsode, if not more.

The early modern discourse of imagination is diffuse. Descriptions of the faculty appear in a range of printed texts and a variety of contexts. Among them, to give just a few examples, are Pierre de la Primaudaye's encyclopedic work of natural philosophy, *Academie françoise*; the physician André du Laurens's study of the eye, *Discours de la conservation de la veue*; the surgeon Ambroise Paré's treatise on birth defects and natural phenomena, *Des monstres et prodiges*; Pierre le Loyer's study of phantoms, *Discours et histoires des spectres*; and the philosophical verse treatises of the poet John Davies. Information about the image-making faculty is to be found in popular medical manuals, travelogues, sermons, dictionaries, and scientific treatises. The psychologists of this period were natural philosophers, supernaturalists, doctors, divines, and poets—many kinds of writers writing in diverse genres, with disparate motivations and for different audiences. The scope of this discourse has been charted recently in Clark's *Vanities of the Eye*, a panoramic cultural history investigating how the notion of visual objectivity was gradually eroded during the early modern period. As Clark shows, unreliable imagination is just one among several contributors to this large-scale effect, in among magic, demons, perspectival art, and skepticism.[15]

The literary implications of imaginative discourse, however, have not received much focused attention. Typically, the cognitive theory of imagination is invoked passingly, in order to contextualize, for example, the representation of fairies or mirrors, or the psychology of Shakespearean protagonists.[16] Arthur F. Kinney's *Shakespeare and Cognition*, which traces the influence of Aristotelian psychology in Shakespearean dramaturgy, is a rare example of a single-author study dealing with the subject; still, its analysis is limited to the invocation of sight and mind by Shakespeare's use of stage properties.[17] Faculty psychology is more often used to parse Renaissance literature, in other words, rather than the other way around. One explanation for the lack of sustained scholarly attention in this area may concern the quality of the discourse. From the perspective of intellectual history, Renaissance faculty psychology is not innovative. The philosophical agenda, Park writes, was "continuous" with thirteenth- and fourteenth-century thought; the sixteenth century only simplified earlier explanations, dispensing with the "accretions and interpolations introduced by medieval commentators" by returning to Aristotle directly.[18] In spite of this, early modern writings on imagination are incoherent at best. Past critics have characterized the discourse as "a chaotic

---

15. See also Yasmin Annabel Haskell's collection of essays dealing with imaginative dysfunction, *Diseases of the Imagination and Imaginary Disease in the Early Modern Period* (Turnhout, Belgium: Brepols, 2011).

16. See, for example, Sean H. McDowell, "Macbeth and the Perils of Conjecture," in *Knowing Shakespeare: Senses, Embodiment and Cognition*, ed. Lowell Gallagher and Shankar Raman (Basingstoke, UK: Palgrave Macmillan, 2010), 30–49; Maurice Hunt, *Shakespeare's Speculative Art* (New York: Palgrave Macmillan, 2011), chap. 1; Judith H. Anderson, "Working Imagination in the Early Modern Period: Donne's Secular and Religious Lyrics and Shakespeare's Hamlet, Macbeth, and Leontes," in *Shakespeare and Donne: Generic Hybrids and the Cultural Imaginary*, ed. Judith H. Anderson and Jennifer C. Vaught (New York: Fordham University Press, 2013), 185–220; and Kevin Pask, *The Fairy Way of Writing: Shakespeare to Tolkien* (Baltimore: Johns Hopkins University Press, 2013), chap. 3.

17. Arthur F. Kinney, *Shakespeare and Cognition: Aristotle's Legacy and Shakespearean Drama* (New York: Routledge, 2006). Werner von Koppenfels likewise offers a cursory consideration of the pathological associations of *phantasia*, seeing Shakespeare's plays as straightforwardly illustrative of prevailing prejudices; see Werner von Koppenfels, "*Laesa Imaginatio*, or Imagination Infected by Passion in Shakespeare's Love Tragedies," in *German Shakespeare Studies at the Turn of the Twenty-First Century*, ed. Christa Jansohn (Newark: University of Delaware Press, 2006), 68–83.

18. Park, "Organic Soul," 476, 479.

jumble of ambiguous or contradictory fact and theory," lacking in "system, thoroughness, and logical consistency," a "mass of conflicting details."[19]

There are good reasons to return to the jumbled discourse of imagination, however, whose foibles we are arguably better placed to appreciate than before. For one thing, given the deepening focus on cognitive matters in Shakespeare studies, a full-length analysis of the imaginative faculty is long overdue. Since the seminal work done on the body by Gail Kern Paster, Michael C. Schoenfeldt, and others, Renaissance literary criticism has become increasingly interested in matters of the mind.[20] Through such topics as madness, the senses, memory, and sympathy, scholars such as Carol Thomas Neely, Lowell Gallagher and Shankar Raman, Evelyn Tribble, and Mary Floyd-Wilson have recuperated historical ideas and practices involving mentality and cognition.[21] Those ideas have been interrogated and shown as anything but primitive or monolithic. Shakespeare and his contemporaries, we now know, did not mindlessly absorb orthodoxies of their time; rather, they negotiated them in dynamic and ingenious ways.

In league with these efforts, in this book I tackle the early modern literature of imagination from a perspective that privileges idiosyncrasy and contradiction over congruence. Like Mary Thomas Crane in *Losing Touch with Nature*, here I register the "ferment, confusion, and angst" of a period when the new ideas that would supplant Aristotelianism were yet "inchoate and in flux," and I consider how literary texts in particular responded to "changing views of the world."[22] Thus, I treat contemporary writings on imagination less as a coherent historical archive than as an archaeological discourse of the kind described by Michel Foucault. Rather than try to reconstitute a "prediscursive"

19. See, respectively, Louise C. Turner Forest, "A Caveat for Critics against Invoking Elizabethan Psychology," *PMLA* 61, no. 3 (1946): 656; Lawrence Babb, *The Elizabethan Malady: A Study of Melancholia in English Literature from 1580 to 1642* (East Lansing: Michigan State College Press, 1951), 67; and Jay Halio, "The Metaphor of Conception and Elizabethan Theories of the Imagination," *Neophilologus* 50, no. 1 (1966): 455.

20. Gail Kern Paster, *The Body Embarrassed: Drama and the Disciplines of Shame in Early Modern England* (Ithaca, NY: Cornell University Press, 1993); Michael C. Schoenfeldt, *Bodies and Selves in Early Modern England: Physiology and Inwardness in Spenser, Shakespeare, Herbert, and Milton* (Cambridge: Cambridge University Press, 1999). See also Katharine Eisaman Maus, *Inwardness and Theater in the English Renaissance* (Chicago: Chicago University Press, 1995); and David Hillman, *Shakespeare's Entrails: Belief, Scepticism and the Interior of the Body* (New York: Palgrave Macmillan, 2007).

21. Carol Thomas Neely, *Distracted Subjects: Madness and Gender in Shakespeare and Early Modern Culture* (Ithaca, NY: Cornell University Press, 2004); Gallagher and Raman, *Knowing Shakespeare*; Evelyn Tribble, *Cognition in the Globe: Attention and Memory in Shakespeare's Theatre* (New York: Palgrave Macmillan, 2011); Mary Floyd-Wilson, *Occult Knowledge, Science, and Gender on the Shakespearean Stage* (Cambridge: Cambridge University Press, 2013).

22. Mary Thomas Crane, *Losing Touch with Nature: Literature and the New Science in Sixteenth-Century England* (Baltimore: Johns Hopkins University Press, 2014), 2–3.

early modern idea of imagination or attempt to regularize mistakes and inconsistencies in the service of recuperating the history of an idea, my approach is to consider the texture and patterns of this discourse for what they are.[23] This book is not an intellectual history of imagination, therefore. More precisely, it is an interpretation of imagination's sixteenth- and early seventeenth-century discursive transmission, focused exclusively on those elements that held Shakespeare's interest.

Some examples will illustrate how attention to literary style rather than philosophical substance reveals the suppleness of Renaissance thinking. Consider La Primaudaye's account, fairly typical in terms of content, which says that imagination works "in the soule as the eye in the bodie, by beholding to receiue the images that are offered vnto it by the outward senses: and therefore it knoweth also the things that are absent, and is amongst the internal senses as it were the mouth of the vessell of memorie. . . . Nowe after that the Imagination hath receiued the images of the senses, . . . then doeth it as it were prepare and digest them."[24] Initially, imagination is the soul's "eye"; a moment later, it is "the mouth" of memory. Interleaved with the visual conceit—ideas as images—is an alimentary metaphor that says the image-making faculty's work is to "prepare and digest" sense impressions, rather as the body absorbs food. Throughout the section from which this passage is drawn, La Primaudaye is also developing a trichotomy of knowledge: knowledge of things present, things absent, and bodiless things; imagination, he says, straddles the first two types. After he concludes his discussion of "imagination," he reflects a little on "fantasy"; a few pages later, he says that the two are effectively the same.[25] This is a hypertaxonomized account, straining simultaneously to situate imagination in relation to the faculties, the senses, the body, God, and knowledge itself. Yet its programmatic ambitions are undercut by its arbitrary presentation and inconclusiveness.

Resemblances and analogies, though they can be adapted to powerful ideological effect, run the risk of evading the author's control. Du Laurens's use of metaphor to stress the hierarchical order among the faculties, for example, ironically conveys the precariousness of that order. He portrays memory and imagination as reason's "handmaides"—"the one to report; the other, to register and write downe"; both "enioy the priuiledges of renowned excellencie" and reside in the brain's "royal palace." In the next sentence, though,

23. Michel Foucault, *The Archaeology of Knowledge*, trans. A. M. Sheridan (London: Routledge, 2002), 47.

24. Pierre de la Primaudaye, *The second part of the French academie*, trans. Thomas Bowes (London, 1594), 146–47. Editions of *Academie françoise* appeared in print starting in 1577.

25. La Primaudaye, 155.

imagination is a royal spy, gathering field reports from the outward senses. In the next, it is a double agent, peddling "vntrue" misinformation to waylay reason, a betrayed "Captaine."[26] A similar slippage occurs in Davies's *Nosce teipsum*, which first describes the soul as a lady of the court, attended by "Phantasie, near handmaid to the mind." Later, however, Phantasie is a magistrate presiding over a different kind of court, weighing "good" against "ill" in her evidentiary "Ballance."[27] The changeable way these authors marshal the handmaid trope indicates the interpretive leeway that existed despite the fact that basic ideas about imagination were relatively stable.

As a final instance, take Pierre Charron's *De la Sagesse*, whose breathless enumeration of imagination's dangers is worth quoting at length: "It makes a man blush, wax pale, tremble, dote, to wauer; . . . it takes away the power and vse of the ingendring parts. . . . Euen in sleepe it satisfieth the amorous desires, yea changeth the sex. . . . Yea it killeth and makes abortiue the fruit within the wombe; it takes away a mans speech, and giues it to him that neuer had it. . . . It taketh away motion, sense, respiration. Thus we see how it worketh in the bodie. Touching the Soule: it makes a man to lose his vnderstanding, his knowledge, iudgement; it turnes him foole and mad-man . . . it inspireth a man with the foreknowledge of things secret and to come . . . , yea it rauisheth with extasies: it killeth not seemingly but in good earnest."[28] Though alarming, the list is not especially unusual: the myriad vulnerabilities of imagination were well known in Renaissance psychology. More noteworthy, perhaps, is the punitive insistence of the rhetoric—the anaphoric accretion that tries to encompass in a single utterance all that imperils the human being. Charron indirectly conveys the challenge of writing about imagination, under whose weight syntax itself seems to groan. Read a different way, this chaotic account follows a certain logic: bodily tremors lead into sexual impotence, sex changes, and abortive gestation. Loss of speech is followed by the loss of "motion, sense, respiration." The collapse of the mind heralds folly, ecstasy, and finally death. This seemingly chaotic, calamitous picture of personal obliteration is also an organized anatomy of imagination, a synopsis not only of fantasy's hazards but also of the structure of human experience.

Early modern faculty psychology is filled with such subtle tensions. Depending on just a few differently chosen words, imagination is akin to judgment, or the antithesis of judgment. It can be proof of man's elegant design

---

26. André du Laurens, *A discourse of the preseruation of the sight*, trans. Richard Surphlet (London, 1599), 73–74. Earlier published as *Discours de la conservation de la veue* (Tours, 1594).

27. John Davies, *Nosce teipsum* (London, 1599), 26, 46–47.

28. Pierre Charron, *Of wisdome*, trans. Samson Lennard (London, 1608), 66–67. Earlier published as *De la Sagesse* (Bordeaux, 1601).

or else the scourge of humanity. It can be partnered with the world of the senses or else the intellective realm. As we will see in this book, imagination's many explicators—church reformers, witch hunters, mathematicians—routinely construe faculty psychology for their own ends, filling in gaps in the theory, improvising definitions, and reorganizing the terminology as it suits them. Routinely, they mobilize literary strategies such as narrative, rhetoric, figure, and irony. To ignore this discursive heterogeneity means to erase what Park and Eckhard Kessler call the "drama and uniqueness" of Renaissance psychology.[29] More importantly, it ignores the epistemological absorptiveness of imaginative discourse. As an idea, imagination is pliable, receptive of external ideas, porously open to intrusions from other systems of thought. This brings me to the first of my two central claims, which is that the sixteenth-century discourse of imagination was directly and indirectly influenced by epistemological and epistemic shifts occurring in protoscientific fields.

I acknowledge that my use of the term *science* is anachronistic. By the medieval Latin *scientia*, the early moderns would likely have understood "any rigorous and certain body of knowledge that could be organized (in precept though not always in practice) in the form of syllogistic demonstrations from self-evident premises."[30] I use the term to refer to theoretical and practical lines of inquiry devoted to garnering objective knowledge about nature—the structure of the body, illness and healing, the material basis of the universe and its governing laws, the plant and animal kingdoms, and methods of knowledge making founded on empiricism and experimentation. It should also be noted that the scientific changes that came about during the early modern period were incremental rather than revolutionary, further slowed by a universal reluctance to reject ancient doctrines. This book heeds the caution of recent historians of science such as Steven Shapin, Margaret J. Osler, and others, who question the supposition that there was a scientific revolution, a "coherent, cataclysmic, and climactic event" by which understandings of the world were "fundamentally and irrevocably changed," or that there was such a thing as a singular or unchanging "scientific method."[31]

---

29. Katharine Park and Eckhard Kessler, "The Concept of Psychology," in Schmitt et al., *Cambridge History of Renaissance Philosophy*, 462.

30. Lorraine Daston and Katharine Park, *Wonders and the Order of Nature, 1150–1750* (Cambridge, MA: MIT Press, 2001), 3.

31. Steven Shapin, *The Scientific Revolution* (Chicago: University of Chicago Press, 1998), 1, 4. See as well Margaret J. Osler, "The Canonical Imperative: Rethinking the Scientific Revolution," in *Rethinking the Scientific Revolution*, ed. Margaret J. Osler (Cambridge: Cambridge University Press, 2000), 3–24; and the introduction and essays in David C. Lindberg and Robert S. Westman, eds., *Reappraisals of the Scientific Revolution* (Cambridge: Cambridge University Press, 1990). For studies emphasizing the revolutionary aspect, see A. C. Crombie, *Augustine to Galileo: The History of Science, A.D. 400–1650*

In a direct epistemological sense, the new science challenged the precepts of faculty psychology. Early scientists distrusted imagination as a method of knowledge making; they would choose increasingly to test presumptions through observation and trial rather than logic and deductive reasoning. Bacon writes in his *Advancement of Learning* that imagination cannot produce *scientia*; if anything, it is the work of scientific epistemes to study and explain the phenomenon of imagination.[32] Reorganizing the disciplines, Bacon links imagination with poetry—a far remove from the privileged position *phantasia* held in scholastic philosophy. Additionally, there was the gathering conviction that not everything conjectured by the ancients was true and that new knowledge had yet to be discovered. Accordingly, natural historians saw the value of gathering information about plants and animals through firsthand experience; physicians sought new ways to distinguish occult phenomena from mental illness; physicists pondered the material basis of the brain's operations. The facticity of faculty psychology was not challenged, exactly—but the new science was not particularly interested in exploring it, either. The anatomist Vesalius, author of *De humani corporis fabrica*, castigated his predecessors for their foolish reliance on imagination, and he did not explain how the faculties fitted in with what he observed of the brain's interior. Comparably, the astronomer Kepler would describe in his work on optics how the eye forms an image on the retina, but he would not address the question of how this image is relayed into thought.

The other, less direct way in which Renaissance science left its mark on faculty psychology was through language. Epistemic concepts creep into the discourse of imagination and come to inflect how Renaissance writers of all stripes describe imagination. The inflection can take the form of novel tropes: with the resurgence of classical atomism, for example, we find figments of imagination compared to atoms and fine particulate matter; similarly, around the time that mathematics demonstrates that the eye works like a lens, mental images are likened to distorted mirror reflections. Another effect is that a long-standing idea in imaginative discourse takes on a novel ambiguity: the notion that a fantastic idea is like a monstrous beast, for instance, becomes complicated by the difficulty of knowing, in an age of marvelous fauna, where the border between naturalness and unnaturalness lies. Similarly, erstwhile com-

(London: Falcon, 1952); A. Rupert Hall, *The Scientific Revolution, 1500–1800: The Formation of the Modern Scientific Attitude* (Boston: Beacon, 1966); Marie Boas Hall, *The Scientific Renaissance, 1450–1630* (New York: Harper and Row, 1966); and Peter Dear, *Revolutionizing the Sciences: European Knowledge and Its Ambitions, 1500–1700* (Princeton, NJ: Princeton University Press, 2009).

32. Francis Bacon, *The Advancement of Learning*, in *The Works of Francis Bacon*, ed. James Spedding, Robert Leslie Ellis, and Douglas Denon Heath (London: Longman, 1857), 3:382–83.

monplace associations, such as the identification of imagination with emptiness and idleness, may be revised in light of alternative thinking about the
substantiality of phantasms and the labor of knowledge making. In the literary realm, existing figures and topoi can take on new life: as scientific consensus moves the seat of cognition from the heart to the brain, for example, the
phenomenology of desire seen in courtly love poetry acquires fresh vigor.[33]

Since its earliest conception, *phantasia* had occupied a liminal position, interceding between sensory perception of the world and the abstract cogita-
tions of the mind, between the illusory and the real, the human and the
superhuman. That it would become additionally involved in the paradigm
shifts of the sixteenth and seventeenth centuries, caught up in philosophy
and science, is perhaps no surprise, no accident—for it seems that the intellectual function of the pre-Cartesian imagination was to accommodate incongru-
ence, bridge rifts, and solve problems. Foucault notes that in between the
ordering codes of culture and the codes of science lies a "third" region where
the arbitrariness of order itself is made manifest; it is at these peripheries of
knowledge that discursive formations become conspicuous as such, "out-
lined precisely where the levels of scientificity and formalization were most
difficult to attain."[34] The Renaissance discourse of imagination, with all its
variation, constitutes an intermediary region of this sort, a place where aesthetic and epistemological presumptions could be recognized as such. The
second of my claims in this book is that Shakespeare's treatment of imagination rests on a recognition of this kind: it draws the disorderly and mediatory
image-making power into the province of art, recharacterizing its endless
generativity as a source of aesthetic creation.

Shakespearean Imagination

In the chapters of this book, I make it plain that Shakespeare thought about
imagination in an impressive range of ways. This thinking is not narrowed by
concerns of aesthetic theory or ethical judgments; rather, it is investigative,
curious about the nature of phantasms and the epistemological implications
of imagination. My readings indicate that Shakespeare was circling a set of
implicit problems, unresolved questions in imaginative discourse. Where in

33. Jenny C. Mann has described how figures of speech can become "vagrant" or disorderly, uti-
lized to accommodate an emerging cultural identity within existing tradition. See *Outlaw Rhetoric:
Figuring Vernacular Eloquence in Shakespeare's England* (Ithaca, NY: Cornell University Press, 2012),
chap. 1.

34. Michel Foucault, *The Order of Things: An Archaeology of the Human Sciences* (New York:
Pantheon, 1970), xx–xxi; Foucault, *Archaeology of Knowledge*, 195.

the body does imagination happen? What causes disorders in perception? What are phantasms made of? Are mental images like ordinary images? What is the role of imagination in nature? Can imagination lead to new discoveries? These questions make it clear that Shakespeare knew imagination to be a multifaceted concept, involved not only with the mind but also with the body, the cosmos, and nature; it is a conceptual cipher, he seems to have understood, under which many ideas were tacitly being worked out. To be sure, Shakespeare was neither the first nor the last to ponder such questions: they were buried in sixteenth-century Renaissance psychology generally, and they drew the attention of scientists, philosophers, and artists alike.

But these are not the type of questions that had been asked by the medievals and ancients, either. Faculty psychology has some of the markers of a scientific paradigm, as defined by Thomas S. Kuhn: it was a mode of solving problems, better than competing modes; it could be used to answer questions considered to be important. As Kuhn notes, at the edges of every paradigm exist problems that it is not well suited to solve.[35] The problems that had occupied philosophers of earlier ages held less fascination for the early moderns. Whereas medieval thinkers had explicitly pondered whether the faculties were ontologically distinct from the soul, had thought with precision about the relations among the faculties, such matters were abandoned by Renaissance commentators. Whereas faculty psychology had once shown how the embodied soul learns about itself, the epistemological legitimacy of imagination was now in question. Whereas imagination had been used before to think about physiognomy, dream interpretation, and memory arts,[36] it would now be called on in the context of anatomy and corpuscular philosophy. The questions raised by Shakespeare's works, then, were firmly contemporaneous, forward looking, even innovative. His thinking goes beyond the original purview of faculty psychology; it rises, rather, to the challenge of adapting that psychology to evolving scientific epistemes and grapples with difficulties and lacunae in scientific knowledge.

In saying this, it is not my intention to suggest that Shakespeare had firsthand knowledge of the new science, that he was abreast of the latest theories and practices, that he read Vesalius and Kepler. I am arguing, rather, that Shakespeare was attuned to the intellectual welter of his time; his engagement with imagination is founded not on scientific thinking so much as on the discursive ripples created within early modern culture by that thinking. What-

---

35. Thomas S. Kuhn, *The Structure of Scientific Revolutions* (Chicago: University of Chicago Press, 2012), 11, 37.

36. Park, "Organic Soul," 469.

ever limited knowledge Shakespeare had of mathematics or medicine, he was demonstrably aware of the patterns of speech, the literary forms, and the generic vogues that these burgeoning epistemes amplified in the sixteenth-century sensibility. As is still true today, philosophy and science in Shakespeare's time had a tendency to creep into the way ordinary people talked about even nonscientific matters. To use such figures and tropes, and comprehend their cultural resonance and significance, it is not necessary to fully understand the domain-specific knowledge that they encode or to know how they came about. Shakespeare's interdiscursive borrowings heighten the aesthetic *energia* of his works, calling attention to what Stephen Greenblatt terms "half-hidden cultural transactions"—those contingent and collective beliefs whose traces are encoded in great literary texts.[37] Though source study suggests that Shakespeare knew of such authors as Le Loyer, La Primaudaye, and Paré—many of the sources cited in this book were available in English translation—my procedure will not be to recuperate Shakespeare's reading habits or speculate about what he personally knew.[38] Rather, I sift the heterogeneous literature of imagination, tracking the incidence of particular rhetorical and literary devices. I identify and contextualize the ubiquitous cultural idioms that evidently caught Shakespeare's ear, and I trace their connections with the epistemological reorientations occurring in Renaissance Europe.

The idea that Shakespeare absorbed and in turn shaped the epistemic ferment of his historical moment through language should come as no surprise, since various studies have already mapped his ability to track and shape movements in culture with his usage of words. Patricia Parker, for instance, has demonstrated that the language of the plays exposes and subverts ruling ideologies using interrogative verbal networks.[39] Douglas Bruster has described the Shakespearean tendency to quote without attribution from a wide range of sources, and Crane has suggested that Shakespeare's brain was acutely apperceptive of polysemic terms and cultural metaphors of all kinds.[40] Shakespeare does not only appropriate or allude; he also adapts, fitting his

---

37. Stephen Greenblatt, *Shakespearean Negotiations: The Circulation of Social Energy in Renaissance England* (Berkeley: University of California Press, 1988), 4.

38. See Geoffrey Bullough, ed., *Narrative and Dramatic Sources of Shakespeare* (London: Routledge, 1957–1975), 7:30, 7:427–28, 7:463; and Kenneth Muir, *The Sources of Shakespeare's Plays* (London: Methuen, 1977), 216. Other authors listed among Shakespeare's sources mentioned in this book are Pierre Boaistuau, Timothy Bright, Ludwig Lavater, John Leo, Pliny, Giambattista della Porta, and Walter Raleigh.

39. Patricia Parker, *Shakespeare from the Margins: Language, Culture, Context* (Chicago: University of Chicago Press, 1996).

40. See Douglas Bruster, *Quoting Shakespeare: Form and Culture in Early Modern Drama* (Lincoln: University of Nebraska Press, 2000), 21, 50; and Mary Thomas Crane, *Shakespeare's Brain: Reading with Cognitive Theory* (Princeton, NJ: Princeton University Press, 2010), 24, 25.

extradiscursive borrowings into existing formal and literary conventions already in place. By being attentive to both linguistic patterning and cultural permeations, I mean to avoid what Stephen Cohen has described as the artificial separation of form and history in Renaissance studies.[41] In attempting to delineate Shakespeare's engagement with imagination, I heed Heather Dubrow's notion of methodological pluralism; I embrace new historicist, new formalist, and historical phenomenological strategies.[42]

As well, *Phantasmatic Shakespeare* participates in the recent critical move to characterize the relation between Renaissance art and science. Several scholars have shown that no clear distinction between C. P. Snow's so-called two cultures existed in the sixteenth century. Elizabeth Spiller points out that scientific work relies on artifice in the form of tools and experimental contrivances; facts are artifacts; the creation and consumption of texts are forms of knowledge practice.[43] Both science and literature, Henry S. Turner observes, drew on a shared substratum of "pre-modern intellectual categories and networks of social relations."[44] Literary studies of this period can challenge the primacy of science as a claim to truth, rather than merely trace the influence of science on culture.[45] This new literary historiography of science importantly exposes the artistic strategies that the scientific enterprise relies on—strategies of performance, persuasion, and representation. More generally, we have begun to acknowledge how culture complements epistemology, filling gaps in knowledge and imposing meaning and coherence on experience: storytelling, Ellen Spolsky says, is an effective means of crafting "good enough" solutions to hard problems; conversely, as Bruce R. Smith notes, the fictions that the early moderns "told themselves about perception, about what was happening in their bodies and brains," reveal a great deal about the felt experience of their

41. Stephen Cohen, introduction to *Shakespeare and Historical Formalism*, ed. Stephen Cohen (Aldershot, UK: Ashgate, 2007), 3.

42. Heather Dubrow, *A Happier Eden: The Politics of Marriage in the Stuart Epithalamion* (Ithaca, NY: Cornell University Press, 1990), 265.

43. Elizabeth Spiller, *Science, Reading, and Renaissance Literature: The Art of Making Knowledge, 1580–1670* (Cambridge: Cambridge University Press, 2004), 25. See also Juliet Cummins and David Burchell, introduction to *Science, Literature and Rhetoric in Early Modern England*, ed. Juliet Cummins and David Burchell (Aldershot, UK: Ashgate, 2007), 2, which notes the role of rhetorical and dramatic techniques in the making of scientific ideas.

44. Henry S. Turner, *The English Renaissance Stage: Geometry, Poetics, and the Practical Spatial Arts, 1580–1630* (New York: Oxford University Press, 2006), vii.

45. See Carla Mazzio, "Shakespeare and Science, c. 1600," *South Central Review* 26 (2009): 4; and Howard Marchitello, *The Machine in the Text: Science and Literature in the Age of Shakespeare and Galileo* (Oxford: Oxford University Press, 2011), 1–3. More essays troubling the presumed distinction between "literary and scientific practices" (xxvi) may be found in *The Palgrave Handbook of Early Modern Literature and Science*, eds. Howard Marchitello and Evelyn Tribble (London: Palgrave Macmillan, 2017).

lives.[46] Shakespeare's plays and poems certainly testify to the problem-solving power of art, in that they translate the epistemic and discursive confrontations between faculty psychology and science into literary convention and form, thus imposing intelligibility on intellectual flux.

The central contention of this book, though, is that Shakespeare does not so much draw on the epistemological tradition of imagination as repurpose the deepening ambiguities of that tradition into the basis of aesthetic representation. The exploration of early modern science that I perform points to the particular challenges that accosted the notion of imagination; from this it is possible to better appreciate Shakespeare's acute perceptiveness as an interpreter of culture, to see how he was tackling the same questions as his proto-scientific peers, sometimes anticipating before his time points that would gain greater visibility in later decades. It is also possible to see the debt that he owed to the many Renaissance thinkers who collectively made the discourse of faculty psychology as variegated and provocative as it was. To suggest that Shakespeare was like a scientist or philosopher would be to disguise his real achievement as an artist, which was, in some sense, to liberate imagination from its epistemological confines and absorb it into the domain of poetry. This absorption does not reduce to yet another form of knowledge making, exactly; if anything, it suggests that art fundamentally differs from science.

As we will see in the upcoming chapters, imagination at the turn of the seventeenth century straddled multiple traditions of thought. It stood at the threshold not only between mind and body, between the human and the divine, but also between aesthetic creation and objective science. By Aristotle, the image-making faculty was originally conceived as a means of explaining how the mind comes to know things in the world. With the destabilizing shifts produced by early modern science, the faculty came to stand increasingly not for knowledge but rather for the limits of knowledge, not for categories but for the transgression of categories, not for an idea of cognition but for the ongoing negotiation of that idea. Arising in an interstitial, intercessional, or disruptive way not just in psychology but also in anatomy, medicine, materialist philosophy, optics, mechanics, and the philosophy of science, imagination had arguably become much more than a mental faculty. For Shakespeare, it was a site of creativity, a site where systematization is frustrated and generative ambiguities bloom instead. Shakespeare's representations of imagination harness this multifarious fecundity, transfiguring philosophical problems into a creative

---

46. Ellen Spolsky, *Satisfying Skepticism: Embodied Knowledge in the Early Modern World* (Aldershot, UK: Ashgate, 2001), 7; Bruce R. Smith, *Phenomenal Shakespeare* (Chichester, UK: Wiley-Blackwell, 2010), 34.

method. In a way, the uncertainties released by the fraying of faculty psychology effectively help account for the psychological incisiveness for which Shakespeare is famous: the confusing idea of imagination calls attention to the mind's fundamental occlusion from itself and underscores the fact that men and women operate always in the absence of complete self-knowledge. When we hear characters like Theseus articulate their perceptual and philosophical notions, we witness directly an inalienable truth about human mentality.

If we accept that Shakespeare's representations draw from the epistemological turmoil of his age, then a reorientation may be in order. For Shakespeare, imagination does not precede representation, is not the inspirational font from which poetry pours. Instead, aesthetic representation is the instrument with which the elusive epistemology of imagination may be understood. We have mistaken the problem of imagination for Shakespeare's power of imagination; we have tended to conflate his engagement with this problem with his personal imaginative skill. The real achievement of the Shakespearean imagination is to recognize in the slipperiness of imagination's epistemological conception a foundation for *poesis*, to make the imagination and its discontents the property of poesy. Leveraging imagination's cognitive rather than aesthetic theorizations, Shakespeare forges a link between mental representation and poetic representation that takes a basically different view from conventional Renaissance aesthetic theory. In Shakespeare, this link is not determined by aesthetic and ethical codes, by the dichotomy of mimesis and fantasy, or by the mysteriousness of human experience. He was certainly aware of those codes—we see them in Theseus's speech, which depicts poetry as a vatic creative transport. But the construction of that speech also suggests that the chaotic pluralism of imagination as an idea was more important to Shakespeare than any singular notion. His instinct was not to quell that pluralism but to unleash it; the objective is to achieve not a centripetal intellectual unification but rather a centrifugal artistic efflorescence.

The chapters of this book have been arranged so as to convey the scope of Shakespeare's interest in mental representation. Though they follow roughly the chronology in which Shakespeare wrote, they do not build in a sequence. Instead, I examine in each a different facet of imagination in conjunction with a different scientific discourse. While the pairings might have been done differently, they have in the main been dictated by the evidence: I have focused on those areas of the canon where the epistemology of imagination manifests itself most provocatively or extensively. To give each of the primary texts its due, I have in general avoided overlap among the chapters. Collectively, these chapters establish that Shakespeare thought about imagination in a number of senses—as a part of the human organism, a type of disorder, a principle in nature, a method, and a metaphor. This array of perspectives not only conveys Shakespeare's lifelong fascination with imagination; it suggests the essentially structural position that imagination held in his thinking, being always a destabilizing force that calls forth creative possibilities no matter the particular context. Finally, a note about terminology: distinctions between such terms as *fancy* and *imagination* were anything but crisp in early modern English. Often, *imagination* and *fantasy* are favored in scientific and philosophical texts, while the abbreviated *fancy* arises more often in literary ones. Too-strong generalizations are ill advised, however, for writers frequently follow their own preferences. Taking my cue from my sources—Shakespeare included—I too use these terms flexibly.

In chapter 1, "Between Heart and Eye," I tie the unclear anatomy of the imagination to the phenomenology of Shakespeare's sonnets. Over the course of its history, the faculty had been linked to different cerebral ventricles, to the senses as well as reason, divided into parts and reunified. In the wake of Vesalian anatomy, physicians faced the additional challenge of reconciling new evidence about the brain's interior with older knowledge. Around this time, the physiological disarray often seen in Elizabethan love poetry reaches a new pitch. In Shakespeare's sonnets, we find the speaker's organs disputing where the beloved's phantasmatic image truly resides; through self-anatomizing corporeal narratives, the lover investigates the connections between his mind and body. Imagination as an epistemological method is explored more fully in the second chapter, "Children of Fancy." Here I argue that the fanciful trifling of *Love's Labor's Lost* deconstructs the gathering view of imagination as a sort of intellectual idling: Bacon and others wrote that imagination is merely a form of childish toying, separate from the systematic labor of discovery. Shakespeare parodically recasts these prejudices by conceiving a pedagogic courtly academia devoted only to fancy, run not by scientists but by fops, pedants, and schoolchildren. In so doing, the play weighs the importance of courtesy, collaboration, novelty, and invention in the making of scientific knowledge.

In religious discourse, imagination was distrusted as a symbol of earthly life's emptiness or vanity. But the very notion of emptiness would be complicated by the atomist philosophies of the sixteenth century, among them the ideas of Lucretius, who posited a universe made up of whirling particles. In chapter 3, "Of Atoms, Air, and Insects," I trace the incidence of materialist tropes in the discourse of imagination, in which phantasms are figured as flies and worms. *Romeo and Juliet*'s imagery of thin substances, the webs and fumes with which the play imagines the particulate texture of dreams and death, culminates memorably in Mercutio's portrait of "Mab." Only through metaphor, suggests Shakespeare, can we grasp the insensible realities of the material cosmos. But what happens to an analogy when the science behind it crumbles? The most intuitive analogue for imagination was the eye: since antiquity, sight and insight had been thought of as contiguous or similar processes. Kepler's optics, however, would reveal that the two are unconnected. As though in response, an idea develops in Renaissance culture that the mind's eye is like a distortive lens. In my fourth chapter, "Seeing to See," I show that Shakespeare saw in the new optical paradigm a framework for representing cataclysm: whereas in *Venus and Adonis* he explores the traumatic severing of eye and mind's eye, in *King Lear* he altogether dismantles the age-old association of blindness and wisdom.

In chapter 5, "Melancholy, Ecstasy, Phantasma," I consider what is perhaps the broadest topic in imaginative discourse: pathology. While imagination and humoral illness had been linked as far back as Galen, in the seventeenth century it seemed that hallucinatory melancholy had reached pandemic proportions. At a time when physicians were trying to distinguish between natural and supernatural disorders, Shakespeare explores the difficulty of diagnosis. Whereas it is possible in *Hamlet* to make deductions about the mind's health, the nebulous mental diseases of *Macbeth* more grimly blur the distinction between normal and pathological ideation. In my final chapter, "Chimeras," I turn to a different sense in which imagination was believed to rule over the body: by engendering monstrous offspring. The power of the fantasy to produce chimeras in the mind is unwittingly mimicked in Renaissance works of natural history, which routinely describe beasts as assemblages of other animals in a way that unconsciously underlines imagination's basic involvement in ordinary perception. *The Tempest*, with its motif of "shape," deliberately highlights what the naturalists miss. The indeterminate forms that populate Prospero's island—Caliban, Miranda, Ferdinand, and others—call into question the supposed distinction between imagination's ability to imitate and to fabricate.

FANCY TRANSFIGURED

Reread in light of the various ways that Shakespeare would mobilize the idea of imagination, Theseus's speech can be seen to be of a piece with the other plays and poems. Though the articulation of fancy in *A Midsummer Night's Dream* may be understood on its own without any invocation of European intellectual history, a fuller understanding of its striking, careful formulations requires knowledge of the lexicon of imagination. That Theseus variously calls phantasms "toys," "shapes," "forms," and "fantasies," for example, is historically accurate—there were indeed many names for mental images in early modern English. The decision to cram them all in here, moreover, is a crucial element of the definition: Shakespeare stresses that no explanation of the image-making faculty can evade its *copia* of tropes, neglect its toylike and shapelike associations, attempt to edit its unruliness. These tropes may well have been more evocative for and intelligible to Shakespeare's auditors than they are for us, as they possess the special edge that certain linguistic formations acquire at a particular cultural moment.

The epistemological efficacy of imagination is broached right at the start: "I never may believe / These antique fables." With "may," Theseus, an Athenian lawmaker, says he is skeptical—*I never could believe*—because he has to be—*I never should believe*. Indeed, "antique fables"—in some editions, "antic"—seems to be pointing to the innumerate outlandish stories told in contemporary medical discourse about imagination's power to delude the ailing and gullible. Theseus's distinction between apprehension and comprehension, meanwhile, suggests he has some knowledge of psychology: first he notes that fantasies "apprehend / More than cool reason ever comprehends"; later, he says that when imagination "would but apprehend some joy, / It comprehends some bringer of that joy." The verbs raise the question of whether imagining should be thought of as a type of perceiving or of reasoning—Albertus Magnus, for one, included imagination among the faculties of "apprehension."[47] Theseus's uncertain thinking on this point conveys the fluid conception of imagination's functions in Shakespeare's time. It voices the worry that uncontrolled imagination can usurp the rule of "cool reason." Spiritual and supernatural associations are invoked too, with "vast hell" and "fine frenzy." The first formulation is crucially ambiguous: when Theseus says that the madman "sees more devils than vast hell can hold," it is unclear whether the devils are really there. Physicians and demonologists of the period thought hard about this nuance, as we will see in chapter 5.[48] A different sort of possession is implied by the oxymoronic "fine frenzy"—a divine furor, in which the poet enters into a rapture, "rauished vnto heauen."[49]

Still, a "phrensie" could also be an inflammation of the brain. Throughout the speech, the realms of the ethereal and aesthetic are interlaced with the cor-

47. Bundy, *Theory of Imagination*, 187. See as well Charron, *Of wisdome*, 54: "To receiue and apprehend the images and kindes of things . . . this is imagination"; and Plutarch, *The philosophie, commonlie called, the morals*, trans. Philemon Holland (London, 1603), 835: the soul "apprehendeth externall things, by the meanes of the imaginative facultie."

48. See, for example, Timothy Bright, *A treatise of melancholie* (London, 1586), 112; and Pierre le Loyer, *A treatise of specters or straunge sights*, trans. Zachary Jones (London, 1605), 34v, 123v. Le Loyer's *Discours des spectres ou visions* appeared in 1586.

49. Le Loyer, *Treatise of specters*, 2r.

poreal and the material; mystical idealism is interwoven with empirical reality. For example, "seething brains" and "cool reason" index the qualities of warmth and moisture that were central to Galenic physic. The humoral aspect of imagination will later arise in *Hamlet*, while the self-deluding "lover" sounds rather like the speaker of the sonnets. Theseus's decision to say "brains" rather than, say, the metrically equivalent "minds" lays a subtle emphasis on the anatomical body. At the same time, his description of poetic inspiration is interestingly framed as an ontological trick: "imagination bodies forth / The forms of things unknown," producing something out of "nothing." "Airy nothing" too conceals a question about the substantiality of mental figments; Shakespeare would explore this question more fully in *Romeo and Juliet*. Similarly, the emphasis here on formation—forms turning into "shapes"; "shaping fantasies"—receives more expanded treatment in the strange animal and human silhouettes of *The Tempest*. "Toys," we will see in chapter 2, evokes the contemporary sentiment that imagination is a sort of purposeless playing, while the repeated conflation of seeing and imagining—the lunatic "sees" demons; the lover "sees" loveliness in his mistress—points to the epistemological unreliability of both.

We have no way of knowing whether these protoscientific nuances were yet in Shakespeare's mind when he composed this speech; quite possibly, it is less a mature summation of Shakespeare's thinking than a developing outline, laying out topics about which he had more to say. As it stands, it can be parsed with the aid of premodern philosophy alone, without invoking scientific ideas: David Schalkwyk and Adam Rzepka have done so, showing that the heterogeneity of Theseus's language affirms the wide compass of imagination's power and the ways that theater trades on and expands that power.[50] My contention, nonetheless, is that even if Theseus is not talking about science, he is not speaking in isolation of scientific thinking either. The work of this book will be to flesh out the cryptic formulations that Shakespeare strews here. For the moment, we can note that the texture of Theseus's speech embodies the richness of imaginative discourse. If its tone has proved difficult for readers to judge, that may be because we are hearing the voices not only of Theseus and Shakespeare but of many other writers and translators too. And if the message is murky, that reflects its subject matter very well too. This definition of imagination is unwilling to explain in a simple way what imagination is because its author was interested in mining, not curtailing, its creative potential.

50. David Schalkwyk, "The Role of Imagination in 'A Midsummer Night's Dream,'" *Theoria* 66 (1986): 51–65; Adam Rzepka, "'How Easy Is a Bush Supposed a Bear?': Differentiating Imaginative Production in *A Midsummer Night's Dream*," *Shakespeare Quarterly* 66, no. 3 (2015): 308–28.

Perhaps because the upshot is not evident, Hippolyta offers a clarifying rejoinder:

> But all the story of the night told over,
> And all their minds transfigured so together,
> More witnesseth than fancy's images
> And grows to something of great constancy;
> But, howsoever, strange and admirable. (5.1.23–27)

Fundamentally, she agrees with Theseus that imagination is "strange." Yet she does not see the caprices of imagination—the mental roving that Theseus has just now described and performed—in the four lovers' "story of the night," which instead "grows to something of great constancy." A. D. Nuttall notes the importance of coherence in Hippolyta's pragmatic epistemology: "If the witnesses separately tell stories that form a consistent whole, they are likely to be telling the truth."[51] She is saying, in a way, that the potential confusions of "fancy's images" can be lessened, or "transfigured," through the intervention of storytelling, of narrative representation. "Transfigure," a word Shakespeare used only once, nicely conjoins language and experience. In one sense, it suggests a transformative melding of multiple minds. In another, it suggests the altering power of figures.[52] We can only tell of imagination by way of figurative language; imagination can therefore be reordered by refiguration. Whereas Theseus said, "I never may believe," Hippolyta's coda is spoken in the passive voice—story told over, minds transfigured—as though to stress the impersonality of communion. Constancy forged through togetherness renders imagination "strange and admirable" in a new sense altogether.

We could also read Hippolyta's lines more allegorically, as a statement of the state of play in early modern imaginative discourse. If Theseus implies that imagination as an idea was being jostled on multiple fronts, Hippolyta indicates that the transfiguring outcome may yet be an affirmative one. Their discussion invites us to compare the merits of intellectual consensus with the more intuitive type of constancy conferred by representation. For the moment, the debate is set aside. Theseus is asked to select the evening entertainment; perhaps still thinking of Hippolyta's remarks, he is drawn to a "tedious brief" scene of "tragical mirth," a mishmash of a play that appears to pose an interesting challenge—"How shall we find the concord of this discord?"

---

51. A. D. Nuttall, *Shakespeare the Thinker* (New Haven, CT: Yale University Press, 2007), 123.

52. Mann argues that the play as a whole is an exercise of rhetorical transfiguration, enmeshing classic and vernacular figures; see *Outlaw Rhetoric*, chap. 6.

(5.1.60). A similar question could be asked of the Renaissance psychology of fancy, caught as it was in the midst of a complicated cultural transformation. In Shakespeare it is not yet possible to see a presentiment of the image-making faculty's post-Cartesian destiny as the pure and sempiternal force of creativity. During Shakespeare's time, poetry was more often conceived in terms of mirroring and making—and Shakespeare does both, albeit with an unusual epistemological bent: his fictions mirror the mutable worlds of early modern knowledge making; they make anew the aesthetic potential of imagination. In the ingenuity of this we must surely recognize some ancestor of imagination as we know it today.

CHAPTER 1

Between Heart and Eye
Anatomies of Imagination in the Sonnets

> Tell me where is fancy bred:
> Or in the heart, or in the head;
> How begot, how nourishèd?
>
> —*The Merchant of Venice*

As Bassanio chooses among three caskets in *The Merchant of Venice*, a song is sung that ponders the location of "fancy" in the body. The two suggested answers are "heart" and "head," but neither is correct: fancy, we are told, "is engender'd in the eye," is "with gazing fed."[1] Ironically, the dramatic context dictates that Bassanio should certainly not trust his eyes, for in order to win the casket game he must select lead in place of gold and silver; in order to arrive at this counterintuitive decision, he would do better to engage his intellect above his senses. In provoking this consideration of heart, head, and eye, the riddle might perhaps have given Shakespeare's audience pause as well: How does "fancy" tie with these three regions of the body? Where in the corporeal anatomy does "fancy" happen? The precise situation of fancy remained an open question in faculty psychology, though it had been considered in various ways. Over the ages, philosophers had debated the number and nature of the different mental faculties of perception: their place in the brain, their relation to the other internal senses, five outward senses, and vital organs. This chapter explores how these questions were complicated by the increasing sophistication and empiricist focus of early modern anatomy,

---

1. William Shakespeare, *The Merchant of Venice*, 3.2.63–68. Shakespeare references are from *The Norton Shakespeare*, ed. Stephen Greenblatt et al. (New York: W. W. Norton, 2016), hereafter cited parenthetically.

exemplified best by the pioneering work done by Andreas Vesalius. Ground-breaking advances in human dissection gave new currency to questions about the physiological underpinnings of mental processes; there lay a challenge in reconciling the philosophy of the inward wits with the documented structure of the body. The upshot is that, in sixteenth-century psychological and medical writings, the anatomical basis for imagination is described in contradictory and ambiguous terms.

These problems were a boon for Elizabethan love poetry. In the tradition of the Petrarchist sonnet especially, the unclear anatomy of imagination demonstrably energizes conventional poetic conceits about love's power to create turmoil in the desiring body. The question of where fancy lives, which surfaces only fleetingly in *The Merchant of Venice*, is investigated more extensively in Shakespeare's sonnets; in them, "fancy"—both erotic desire and imaginative perception—appears to circulate among different bodily locales. We repeatedly find the speaking lover tracing the corporeal presence of his beloved's phantasmatic image as it drifts through his being, disrupting or else reconfiguring the usual relations between his heart, eye, and mind. The lover contemplates the bodily effects activated by this inward image—malfunctioning organs; fleeting but potent impressions of the physiological force of mental imagery; and the interdependent movements of passion and thought. In so doing, he demonstrates how the hunt for the embodied imagination proliferates narratives about the body's workings. In these narratives, the speaker of Shakespeare's sonnets grapples with the same kinds of problems that psychologists and physicians of the period were negotiating: the connection between the body and mind, the relation between corporeal form and physiological function, and the methodological advantages of philosophical speculation versus empirical observation. Though he does not conclusively identify fancy's breeding ground, Shakespeare's lover approaches the kind of humanistic self-knowledge that was in fact the objective of early modern anatomy.

ANATOMIES OF IMAGINATION

When Aristotelian ideas about the soul came to be refracted through Islamic philosophy and then medieval scholasticism, they became grounded in the body. Before the Renaissance, there had been much speculation already about the corporeal embodiment of the mental faculties, their precise arrangement in the head, and their exact number and particular functions. This discussion continues in the anatomical, medical, and psychological literature of the sixteenth and early seventeenth centuries; the authors of these texts continue to weigh the finer points that had engaged earlier philosophers. The extant problems, moreover, had now to be reconciled with anatomical evidence of the brain's interior, evidence growing not merely in quantity, thanks to the work of Vesalius and others, but also in epistemological legitimacy. That this evidence neither clearly corroborated nor refuted the idea of a faculty-based mind is a difficulty that contemporary accounts of the anatomical imagination handle in various ways, through self-conscious hedging, syncretic contrivances, and ideological postures of confidence.

Early modern psychological commentaries are inconsistent as to how many faculties there are and what they each do. The mind's perceptual apparatus proves the most problematic. At most, there are three faculties involved in the work of mental representation: the common sense, or *sensus communis*, which consolidates the streams of impression gleaned from the five senses into one; the fantasy; and the imagination. Sometimes, these three are reduced to two or just one faculty.[2] Different authors make the distinctions differently, and with different levels of discrimination. Thomas Vicary, surgeon to Henry VIII, separates "fantasy" from "imagination": the former "taketh all the formes or ordinances that be disposed of the fiue Wittes," while the latter receives "the fourme or shape of sensitiue things."[3] Somewhat confusingly, the physician Philip Moore lists the three mental faculties as "imaginacion or common sense," "reason or phantasie," and "memorie"—he calls reason by the name "phantasie," as medieval thinkers had sometimes done.[4] Vestiges of medieval faculty psychology can be seen as well in the thinking of John Jones, a physician and translator of Galen's *Elements*: he refers to the perceptual powers as "apprehension, fantasie, imagination, opinion, and common sense."[5] The terminological discrepancies are so severe that the poet John Davies of Hereford declares his intention to clear things up:

> *Imagination, Fancie, Common-sence,*
> In nature brooketh oddes or vnion,
> Some makes them one, and some makes difference,
> But wee will vse them with distinction.
> With sence to shunne the *Sence* confusion.[6]

---

2. See Pierre de la Primaudaye, *The second part of the French academie*, trans. Thomas Bowes (London, 1594), 131. La Primaudaye's *Academie françoise* appeared first in 1577.

3. Thomas Vicary, *The Englishemans treasure* (London, 1587), 15–16.

4. Philip Moore, *The hope of health* (London, 1565), 8v.

5. John Jones, *The arte and science of preseruing bodie and soule* (London, 1579), 104.

6. John Davies of Hereford, *Mirum in modum* (London, 1602), B3r.

Whether Davies succeeds in dispelling "the *Sence* confusion" is arguable: he does clarify that the job of the *"Common-sence"* is to relay phantasms to *"Fantacie,"* but does not explain what differentiates the *"Imagination."*[7] It is easy to see why Pierre de la Primaudaye, author of *The French academie*, decided that it is permissible to use the terms "indifferently"; he points out that *"Fantasie* is deriued from a Greeke worde that signifieth as much as *Imagination."*[8]

The precise number of faculties had ramifications for the presumed architecture of the brain, for it was commonly believed that each of the faculties resided in a particular cell or ventricle of the brain. And the physical arrangement of the cerebral *cellulae* reflects the relation between the different powers. Accordingly, the surgeon Ambroise Paré writes that there are four ventricles—two in the front, one in the middle, and one at the back: "The most proper benefit of the two first ventricles of the braine is to entertaine the Phantasie as in a convenient seat and habitation, seeing the minde there estimates and disposes in order the species of things brought in from the externall senses, that so it may receive a true judgement of them from reason which resides in the middle ventricle."[9] Imagination is in the anterior regions because it works closely with the eyes, ears, nose, and mouth. Conversely, memory lies in the hindermost ventricle, being the brain's "Treasury and store-house." In other words, the faculties are arrayed, front to back, according to their distance from the physical world. This idea, though appealing in its neatness, was impossible to verify. As the physician Juan Huarte admits, it is difficult "to know in which of these ventricles the vnderstanding is placed, in which the memorie, and in which the imagination, for they are so vnited and nere neighboured, that . . . they [cannot] be distinguished or discerned."[10] Given that imagination, understanding, and memory are "very different operations," Huarte writes, one might expect each to possess its own corporeal "instrument." But cerebral dissection does not bear this out: "If we open by skill, and make an anatomie of the braine, we shall find the whole compounded after one maner, of one kind of substance, and like, without parts of other kinds, or a different sort."[11] There is another reason why Huarte is

---

7. Freely mixed throughout Davies's text are Aristotelian categories (the motive, sensitive, and principal powers), physical brain tissue (the *pia mater* and *dura mater*), and abstractions such as spirit, contemplation, and truth. Davies, *Mirum in modum*, B1r, B2r, B3v, E1r, D1v.

8. La Primaudaye, *Second part of the French academie*, 155.

9. Ambroise Paré, *The workes of that famous chirurgion Ambrose Parey*, trans. Thomas Johnson (London, 1634), 168. Paré's collected works were earlier published as *Les oeuvres de M. Ambroise Paré* (Paris, 1575).

10. Juan Huarte, *The examination of mens wits*, trans. Richard Carew (London, 1594), 54. Published as *Examen de ingenios* in 1575.

11. Huarte, 52.

inclined to believe that the faculties are not localized in particular sections of the brain: "Considering that the vnderstanding cannot worke without the memorie be present, . . . and that the memorie cannot do it, if the imagination do not accompany the same . . . we shall easily understand, that all the powers are vnited in euery seuerall ventricle, and that the vnderstanding is not solely in the one, nor the memory solely in the other, nor the imagination in the third, as the vulgar Philosophers haue imagined."[12] This explains why damage suffered by even one ventricle translates to a "great abatement" in the operations of all three faculties: "By the annoiance of any one, all the three are weakened."[13]

As Huarte and others knew, the Greeks had not subscribed to this notion of ventricular separation; it had been developed by later Islamic philosophers. While Renaissance anatomists generally deferred to Galen, they were reluctant to dispense with Avicenna's intuitive, elegant model. The physician André du Laurens, for example, faithfully lists the reasons why the ancients believed the faculties "be al of them in all and euery part of the braine"; yet, he defends the "goodly reasons" of the early medievals with greater vigor: "It is very certaine, that are diuerse pettie chambers in the braine, which the Anatomists call Ventricles, these chambers are not for nothing, yea and there is no man that can thinke, that they were made for any other vse, then as lodgings for [the] three faculties." It is reasonable to assume that imagination inhabits the frontal ventricles: "The appearance of the truth of this thing is very great." The study of physiognomy, Du Laurens writes, has established that variations in the shape of the head correlate to differences in cognitive ability. Furthermore, certain well-documented medical disorders—amnesia, for example— more or less prove that injuries to one faculty do not necessarily affect the others. In the end, Du Laurens concludes that the Greeks were probably right. But he adds, remarkably, that theirs is the less intuitive view and therefore likely to be the less influential: "That of the Arabians will euer be more followed of the common people, for that it hath in it a greater shew of euident cleerenes."[14]

The most important work of early modern anatomy did not attempt to reconcile faculty psychology with scientific findings. Vesalius's *De humani corporis fabrica* of 1543 surpassed premodern studies of the human body, documenting the body's skeletal and muscular fabric in exquisite detail, translating

---

12. Huarte, 54–55.

13. Huarte, 56.

14. André du Laurens, *A discourse of the preseruation of the sight*, trans. Richard Surphlet (London, 1599), 78, 80. Earlier published as *Discours de la conservation de la veue* (Tours, 1594). Paré says something very similar, confessing that he prefers the opinion of "the Arabian Physitions," though he does not say why; see his *Workes*, 169.

familiar organs such as the eye, brain, and heart into complex, three-dimensional webs of tunics and membranes, by way of text as well as image. It announces a commitment to empirical observation and firsthand scrutiny and rejects imaginative conjecture. What the *Fabrica* did not do was lay to rest the mysteries of perception and thought that had busied philosophers for ages. Vesalius admitted that, to an extent, "I am able to understand the functions of the brain in animal vivisections with considerable probability and more or less accurately. But I do not have a satisfactory grasp of how the brain performs its functions in imagination, reasoning, thought, or memory (or however else you might like to subdivide or enumerate the powers of the chief spirit in accordance with anyone's theories)."[15] Thus, the most illustrious anatomist of the Renaissance does not explain, or pretend to understand, how the major mental operations—"imagination, reasoning, thought, or memory"—are carried out. Though Vesalius' drawings of the brain show the ventricles laterally arranged side by side, the medieval arrangement was linear, the ventricles stacked one behind the other (figure 1).[16] Vesalius had himself been taught this "scholastic diagramming" of the ventricles, and he thought it laughable: "These are the fantasies of people who never look at the inventiveness of our Maker in the fabric of the human body."[17] It probably would have irritated him that plagiarized illustrations from the *Fabrica* were reproduced in later anatomical reference works: his dissected heads appear, for example, in Thomas Johnson's edition of Paré's works, as well as the *Mikrokosmographia*, the English anatomical compendium compiled by James I's physician Helkiah Crooke (figure 2). In these texts, descriptions of the premodern ventricular model cohabit with Vesalius's carefully annotated diagrams—the former are not clearly connected to the latter. The juxtapositions render all the more conspicuous the synthetic attempts made by contemporary physicians to reconcile faculty psychology and gross anatomy.

Each section of the *Mikrokosmographia* is split into two parts, the first detailing the corporeal circuitry with diagrams and nomenclature, while the second, entitled "Controversies," explains how everything works, offering to resolve such questions as "Whether the Braine be the seate of the Principall Faculties." The book's textual organization illustrates two contrasting ap-

---

15. Andreas Vesalius, *The Fabric of the Human Body*, trans. and ed. Daniel H. Garrison and Malcolm H. Hast, 2 vols. (Basel, Switzerland: Karger, 2014), 2:1258. I have relied on this translation, as well as the Latin facsimile edition, *De humani corporis fabrica libri septem* (Brussels: Culture and Civilization, 1964).

16. On the medieval anatomical arrangement of the faculties, see Simon Kemp, *Cognitive Psychology in the Middle Ages* (Westport, CT: Greenwood, 1996), 51.

17. Vesalius, *Fabric of the Human*, 2:1259.

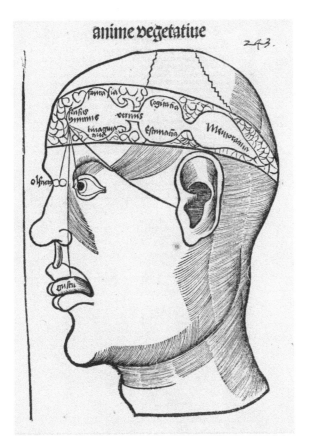

FIGURE 1.   The mental faculties as arranged in the brain, including, from left to right, *sensus communis*, *fantasia*, *imaginativa*, *cogitativa*, *estimativa*, and *memorativa*. From Gregor Reisch, *Margarita philosophica nova* (Strasbourg, 1508), 243. Wellcome Library, London.

proaches to the study of the body, both of which Renaissance anatomists pursued well into the seventeenth century: one focuses on the architecture of the human organism, while the other concentrates on how the soul executes its offices. Andrew Cunningham notes that early modern anatomy, *"anatomy as structure,"* was principally concerned with learning about the body through dissection and inspection of the viscera. *"Anatomy as key to function,"* in contrast, is concerned with how bodily operations are carried out.[18] Plato and Aristotle had developed functional theories, whereas Galen

18. Andrew Cunningham, *The Anatomical Renaissance: The Resurrection of the Anatomical Projects of the Ancients* (Aldershot, UK: Scolar, 1997), 29.

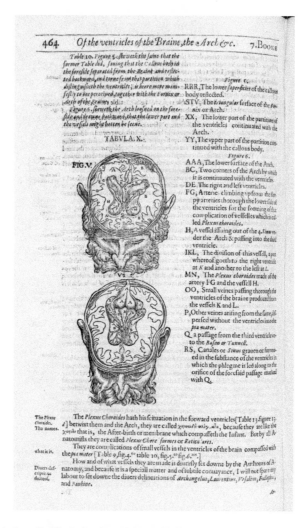

FIGURE 2.   Andreas Vesalius's drawings of the cerebral ventricles, reproduced in Helkiah Crooke's *Mikrokosmographia*. From Helkiah Crooke, *Mikrokosmographia* (London, 1615), 464. RB 53894, Huntington Library, San Marino, California.

and Avicenna had turned their attention to the physical fabric.[19] These two lines of inquiry were not incompatible, Nancy G. Siraisi confirms: there was no "radical or unambiguous separation of the medical from the philosophi-

19. See Kemp, *Cognitive Psychology*, 38: For Aristotle, the mind has no "physical organ" and is "nothing but a kind of capacity for reception."

cal approach to physiology"; differences between Aristotle and Galen were not settled by dissection.[20]

The appeal of synthesizing these two methodologies was evidently strong and not soon to subside. Take, as a final example, this paraphrase of René Descartes's account of vision, written by the English natural philosopher Kenelm Digby: "He conceiueth light to be nothing else but a percussion made by the illuminant vpon the ayre . . . that striketh also our sense; which he calleth the nerue that reacheth from the place stricken (to witt, from the bottome of our eye) vnto the braine. . . . The part of the braine which is thus struken, he supposeth to be the fantasie, where he deemeth the soule doth reside; and thereby tekth notice of the motion and object that are without."[21] Wonderfully brought together in this passage are the material, physiological, psychological, and metaphysical realms: an *atomic* percussion activates an *anatomical* nerve, which alerts the relevant *faculty* and so invokes the *soul*. Digby's account seems almost wishful, articulating a hope that the accumulated knowledge of the centuries may yet be brought together, magnificently bound up, with nothing lost or relegated to inconsequence. The competition among systems of thought was arguably even more acute in the sixteenth century: at its start, the Neoplatonist Gianfrancesco Pico della Mirandola glossed over the question of anatomy in his treatise on imagination; the question is too difficult, "since the imagination itself is midway between incorporeal and corporeal," and at any rate not useful.[22] Around a hundred years later, Francis Bacon saw the question as crucial because it could unlock the "concordances between the mind and body," and because prior attempts to answer it had been sloppy and full of errors.[23] It is perhaps no exaggeration to say that the anatomy of imagination was a matter of implicit paradigmatic importance in early modern science.

As Du Laurens, Huarte, and Crooke separately illustrate, faculty psychology had not been ousted by empirical anatomy. Notwithstanding observable data about the brain's interior, there remained a need for coherent accounts of how the mind performs its actions; if anything, the project of reconciling older models of psychology with emerging knowledge had become more pressing than ever before. The early modern enterprise of locating fancy

20. Nancy G. Siraisi, *Medieval and Early Renaissance Medicine: An Introduction to Knowledge and Practice* (Chicago: University of Chicago Press, 1990), 81.

21. Kenelm Digby, *Two treatises in the one of which the nature of bodies, in the other, the nature of mans soule is looked into* (Paris, 1644), 275–76.

22. Gianfrancesco Pico della Mirandola, *On the Imagination*, trans. and ed. Harry Caplan (New Haven, CT: Yale University Press, 1930), 37.

23. Francis Bacon, *Of the Dignity and Advancement of Learning*, in *The Works of Francis Bacon*, ed. James Spedding, Robert Leslie Ellis, and Douglas Denon Heath (London: Longman, 1857), 4: 378.

meant, in effect, trying to work out how the sensitive soul, taxonomized authoritatively over countless years, maps onto an anatomical body whose features were coming into increasingly sharp focus. The image-making faculty, precisely because of its positional instability, prompted Renaissance psychologists to perform new argumentative moves: their anatomies of imagination, so to speak, are conjectural and inconclusive, made up of strategic approximations and synthetic claims. Viewed another way, the fluid psychology of fancy apparently created a space for thinking through questions about the body and mind and how the two fit together. Straddling the corporeal and the mental, mediating between sense and reason, imagination is a symbol for liminality, an emblem of the complexity of human physiology, the mystery of embodiment.

Between Heart and Eye

Though fancy may have posed problems for psychologists, it furnished literary practitioners, particularly Elizabethan sonneteers, with artistic possibilities. This is not surprising if we consider that the sequence of phantasm creation is not very different from the inception of erotic love as traditionally imagined in the canon. In the usual progression of faculty psychology, sense impressions provoke a response within the body's perceptual apparatus; as a consequence, an image is created in the mind, which in turn comes to shape how future stimuli will be interpreted. Not dissimilarly, eros enters through the eye, swiftly penetrates the heart, and there installs an inward idol whose existence colors subsequent experience. The physiological mobility of the phantasm, we will see, dovetails very well with the corporeal and psychical disarray featured in Petrarchist love poetry, with its jurisdictional quarrels among bickering organs, synesthetic confusion among the senses, and plaintive bargaining between pining lover and unruly body.

In Petrarch's *Canzoniere*, the two organs typically upset by love are the heart and eye. In rima 3, for example, the lover describes his defenselessness before his lady thus: "Love found me altogether disarmed, and the way open through my eyes to my heart."[24] In rima 84, the lover laments, "Eyes, weep; accompany the heart, which suffers death through your fault."[25] At times, eyes and heart conflict with one another, both being besieged by the beloved; at others,

---

24. Petrarch, *Petrarch's Lyric Poems*, trans. and ed. Robert Durling (Cambridge, MA: Harvard University Press, 1976), 38. The original lines are as follows: "Trovommi Amor del tutto disarmato, / et aperta la via per gli occhi al core" (39).

25. Petrarch, 186–87. The original line is "Occhi, piangete, accompagnate il core che di vostro fallir morte sostene."

the lover's body is disrupted by the manifestation of her image. Rima 94 describes how a visual impression quickly morphs into a mental image that is then absorbed into the body, impacting its different parts: "When through my eyes to my deepest heart comes the image that masters me, every other departs, and the powers that the soul distributes leave the members an almost immobile weight."[26] Petrarch's use of *imagine*—a word carrying both a pictorial and a phantasmatic sense—cements the visual quality of inward representation. In rima 83, we are told that the lover's sleep is disturbed by love's "harsh cruel image" (*l'imagine aspra et cruda*). In rima 107, "love-inspiring rays" multiply the internal images so widely (*l'imagine lor son sì cosparte*) that the speaker finds himself bedazzled by them.[27] Strife between heart and eye, and other regions of the body too, was a favorite topos among Elizabethan poets.[28] As in Petrarch, inward images recur in English love poetry, often carrying metaphysical and religious connotations. In *Astrophil and Stella 5*, for example, Philip Sidney writes that

> what we call Cupid's dart
> An image is, which for ourselves we carve;
> And, fools, adore in temple of our heart.[29]

Likewise, in *Amoretti 45*, Edmund Spenser's lover exhorts his lady to give up her "Christall clene" glass and instead use his heart for a mirror, for in it "the fayre Idea of your celestiall hew . . . remains immortally."[30] Both Sidney's "temple" and Spenser's "Idea" evoke Platonist ideality, as well as the concept of Christ's picture inscribed in the believer's breast.

---

26. Petrarch, 196. The original reads,
Quando giugne per gli occhi al cor profondo
l'imagin donna, ogni altra indi si parte,
et le vertù che l'anima comparte
lascian le membra quasi immobile pondo. (197)

27. Petrarch, 186–87, 214–17.

28. In addition to the examples discussed in this chapter, see, for example, Thomas Watson, *The tears of fancie* (London, 1593), B3v, "My hart accus'd mine eies and was offended" (sonnet 20); Henry Constable, *Diana* (London, 1592), 2v, "My Reason absent, did mine eyes require" (sonnet 12); John Davies, *The Gulling Sonnets* (New York: Columbia University Press, 1941), "Mine eye, mine eare, my will, my witt, my harte" (sonnet 5). Outside the sonnet tradition, see Richard Edwards, *The paradise of daintie deuises* (London, 1585), H3v–H4r, "A Dialogue betweene the Authour and his Eye"; Nicholas Breton, *Brittons bowre of delights* (London, 1591), B4v, "Mine eyes haue seene the Idoll of my heart."

29. Philip Sidney, *Astrophil and Stella*, in *The Major Works*, ed. Katherine Duncan-Jones (Oxford: Oxford University Press, 2009), 154.

30. Edmund Spenser, *Amoretti and Epithalamion*, ed. Kenneth J. Larsen, Medieval and Renaissance Texts and Studies 146 (Tempe: Arizona State University, 1997), 86.

At the same time, heart and eye frequently deal with a third entity— "fancy"—whose presence is typically attended by affective turbulence. In Thomas Watson's *Tears of fancie*, heart imposes a "penance" on eyes, which is that they "should weepe till loue and fancie dies": "fancy" is a disruption from which one organ barricades itself from the other.[31] In Robert Tofte's *Alba*, fancy is sandwiched in between heart and eye in a sequence of blame laying:

> My Soule vpon my Hart for this doth plaine,
> My Hart (againe) my Fancie doth accuse:
> My Fancie saith, mine Eyes were too too blame,
> Their ouer-boldnes wrought this great Abuse.[32]

The castigating tone is worth noting: the pathos of emotional turmoil modeled by Petrarch becomes intensified in the Elizabethan idiom into a more convoluted, more agitated expression of phenomenological volatility. In Barnabe Barnes's *Parthenophil and Parthenophe*, this agitation is incendiary, heart becoming engulfed in fancy's flame:

> Through mine eye, thine eyes fier inflames my lyuer,
> Mine eyes in hart, thine eyes cleare fancies write:
> Thus is thine eye to me my fancies giuer,
> Which from thine eyes, to mine eyes take their flight,
> Then pearce the secret center of my harte,
> And feede my fancies with inflamed fewell.[33]

The situation is not novel: the lady's gaze transmits "fancies" to the speaker, which kindles his liver, pierces his heart, and fuels new "fancies" in him. But the poetry is a dense verbal tangle of "thine" and "mine," heart and eyes; the obfuscations of the language mimic the permeating power of fancy, which travels through air, drifts through bodily barriers, and replicates itself. In the final couplet, "eyes, and fancies" are "blind and raging" yet also exultant.

These poets exploit ambiguities in early modern cognitive theory to articulate the cognitive experience of erotic obsession. *Fancy* plays a key role in these representations, for it stands not only for love but also for the idea of love, love's phantasmatic manifestation in the mind; it denotes emotion as well as ideation. The fugitive meaning of the word mirrors the elusiveness of the

---

31. Watson, *Tears of fancie*, B3v.
32. Robert Tofte, *Alba* (London, 1598), B7v.
33. Barnabe Barnes, *Parthenophil and Parthenophe* (London, 1593), 60–61.

experience. The dynamism of desire is thus intensified by the integration of imagination into the overall physiological equation. As one last example of the totally fundamental though muted presence of imagination in the Petrarchist sonnet, take the thirty-third sonnet of Michael Drayton's *Idea*, which bears the dedicatory title "To Imagination":

> Whilst yet mine eyes do surfeit with delight,
> My woeful heart, imprisoned in my breast,
> Wisheth to be transformèd to my sight,
> That it, like those, by looking might be blest;
> But whilst my eyes thus greedily do gaze,
> Finding their objects over-soon depart,
> These now the other's happiness do praise,
> Wishing themselves that they had been my heart,
> That eyes were heart, or that the heart were eyes,
> As covetous the other's use to have;
> But finding Nature their request denies,
> This to each other mutually they crave:
> That since the one cannot the other be,
> That eyes could think, or that my heart could see.[34]

Synesthetic envy is the conceit; heart craves the sensory pleasures of sight, while eye longs for the permanence of heart's delight. Unable to change places, each yearns to be the other: "That eyes could think, or that my heart could see." What exactly makes this a poem about "imagination"? Is it that its speaker is represented in the act of imagining his beloved? Is the point that poetry is a supreme expression of imagination? Or maybe it is that the speaker's introspection, his self-anatomizing, can be done only by imagining what is occurring inside him? All of these meanings would seem to be in play: *imagination* names a range of phantasmatic acts that attend the translation of desire into poetic representation.

Shakespeare's sonnets do not mention *fancy* or *imagination* either. But this does not mean that the image-making faculty is absent from these poems. The dialogues involving heart, mind, and eye that are repeatedly staged in these sonnets interrogate with precision how perception and cognition relate to volition. In a series of distinct contemplative exercises, the speaking lover

---

34. Michael Drayton, *Englands heroicall epistles* (London, 1599), Q2. Drayton dedicated other sonnets to the soul, sleep, and the senses, thus broadening the thematic focus of his sequence beyond the Platonic "Idea." See Joseph A. Berthelot's discussion in *Michael Drayton* (New York: Twayne, 1967), 31.

examines the anatomical movements of his fancy again and again. The multiplicity of these narratives is part of the point, as though to say that it is ultimately more important to ask the question of where fancy resides than to settle on an answer. The poems that follow illustrate how the hunt for the physiological imagination is a quest for a means of representing the body in satisfying ways, and that imaginative representations—that is to say, those kinds of phantasms that are produced by poetry—play a crucial role in this quest.

That Shakespeare knew about the mental faculties is clear from sonnet 141, which refers to five "inward" wits correspondent to the five outward senses:

> In faith, I do not love thee with mine eyes,
> For they in thee a thousand errors note,
> But 'tis my heart that loves what they despise,
> Who in despite of view is pleased to dote;
> Nor are mine ears with thy tongue's tune delighted,
> Nor tender feeling to base touches prone,
> Nor taste, nor smell, desire to be invited
> To any sensual feast with thee alone;
> But my five wits nor my five senses can
> Dissuade one foolish heart from serving thee,
> Who leaves unswayed the likeness of a man,
> Thy proud heart's slave and vassal wretch to be.
> > Only my plague thus far I count my gain,
> > That she that makes me sin awards me pain.

Ultimately, neither the mind nor the body is able to challenge the tyranny of the adoring heart, whose unflinching authority can be heard in the internal rhyme in "dissuade" and "unswayed." Other organs are attached to passive verbs: "delighted," "prone," "invited." Enumerating the senses one by one, the poem reduces the speaker to his eyes, ears, and tongue; the fragmentation is different from the customary partitioning of the *blazon*, in which the lady is segmented into the components of her beauty—hair, cheeks, neck, and so on. Here we see the confounding of corporeal functions rather than bodily parts; the body in question, moreover, belongs to the lover, not the love object. Though the speaker lists several parts of himself, eyes and heart are subtly singled out in a special relationship, tied together by the first end rhyme: "In faith I do not love thee with mine eyes, . . . But 'tis my heart that loves what they despise."

The interplay between heart and eye, and the conceit of physiological conflict, is more fully developed in sonnets 46 and 47. In them, the phantasm
emerges as the necessary means of resolving divisions within the self. Here is
sonnet 46:

> Mine eye and heart are at a mortal war
> How to divide the conquest of thy sight.
> Mine eye my heart thy picture's sight would bar,
> My heart mine eye the freedom of that right.
> My heart doth plead that thou in him dost lie
> (A closet never pierced with crystal eyes),
> But the defendant doth that plea deny
> And says in him thy fair appearance lies.
> To 'cide this title is impanelèd
> A quest of thoughts, all tenants to the heart,
> And by their verdict is determinèd
> The clear eye's moiety and the dear heart's part,
> > As thus: mine eye's due is thy outward part,
> > And my heart's right, thy inward love of heart.

The confusing syntax mimics the difficulty of determining precedence between the two organs—"mine eye my heart"; "my heart mine eye." There is
no "I" in this lyric (unless we count the homophone "eye"). Again, the speaker
is intrigued more by the internal logic of his body than his interpersonal relationship with the beloved. The diction underlines territory and transgression—
"bar," "freedom," "tenants." The poem's central conceit is a mock trial, complete
with defendant, jury, and verdict; in the second quatrain, "plead" turns into
the more juridical "plea." And yet, this trial or "quest" seems a bit one-sided:
the "impanelèd" jurors are the heart's own tenant "thoughts." Why is this a
stacked jury? And how can a heart have "thoughts"?

For centuries, the heart had been the seat of cognition. In the Bible, the
organ represents an array of powers including imagination, perception,
reason, conscience, and piety.[35] As Scott Manning Stevens describes, the ideological

---

35. See, for example, Matthew 9:4, "And Jesus knowing their thoughts said, Wherefore think ye
evil in your hearts?"; Romans 1:21, "When they knew God, they glorified *him* not as God, neither
were thankful; but became vain in their imaginations, and their foolish heart was darkened"; Luke
24:38, "And he said unto them, Why are ye troubled? and why do thoughts arise in your hearts?"; and
Luke 1:51, "He hath shewed strength with his arm; he hath scattered the proud in the imagination of
their hearts." See *The Bible: Authorized King James Version with Apocrypha*, ed. Robert Carroll and Stephen Prickett (Oxford: Oxford University Press, 1997).

work done by the heart is varied: in liturgical practice and in the visual imagery of the Catholic Church, for example, it is a recurring metaphor for the
self.[36] As William E. Slights has shown, the heart in the age of Shakespeare
was embedded in theological, anatomical, and medical discourses, bound up
with ideas not just of the body but also of the soul, faith, narrative, and art.[37]
In the earlier half of the sixteenth century, Erasmus was still able to write, "The
seate of the soule or minde, is in the heart."[38] And Thomas Elyot, diplomat
and physician, wrote this definition in his dictionary: "*Cor, cordis*, the herte,
somtyme it is taken for the mynde."[39] That being said, in Renaissance anatomical literature the heart's physiological role was no longer central in the way it
had been in classical sources. The gathering consensus was that "the principall seate of the soule is in the braine," not the heart.[40]

For this reason, Shakespeare's decision to imbue the heart with implicit but
muted power in sonnet 46 is shrewd: from the perspective of early modern
discourse, this organ represents the undecided locus of the body's intellective
center. Though generally a metonym for romantic devotion, the "heart" of
Shakespeare's sonnets takes many guises. It is a source of moral corruption in
sonnet 62, infecting "all mine eye" and "all my soul" with sinful love. Elsewhere, it stands for the power of imagination:

Those parts of thee that the world's eye doth view
Want nothing that the thought of hearts can mend. (Sonnet 69)

In sonnet 93, it stands for the inaccessibility of others: "Thy looks [are] with
me, thy heart in other place." In sonnet 119, it is a sign not of judgment but
of the erosion of judgment:

What wretched errors hath my heart committed
Whilst it hath thought itself so blessèd never?

---

36. Scott Manning Stevens, "Sacred Heart and Secular Brain," in *The Body in Parts: Fantasies of Corporeality in Early Modern Europe*, ed. David Hillman and Carla Mazzio (New York: Routledge, 1997), 263–84.

37. William E. Slights, *The Heart in the Age of Shakespeare* (Cambridge: Cambridge University Press, 2008).

38. Desiderius Erasmus, *The first tome or volume of the Paraphrase of Erasmus vpon the newe testamente*, trans. Nicholas Udall (London, 1548), L2r.

39. Thomas Elyot, *The dictionary of syr Thomas Eliot knyght* (London, 1538), E2r.

40. Du Laurens, *Discourse*, 3. See as well Helkiah Crooke, *Mikrokosmographia* (London, 1615), 367, which describes the heart as "the root of the Arteries and Author of the Pulse." As its extended title indicates, Crooke's work is a compilation of the work of sixteenth-century anatomists: "Collected and translated out of all the best authors of anatomy, especially out of Gasper Bauhinus and Andreas Laurentius."

This range of connotations underlines the heart's unfixed significance, its plural physiological agency—the fact that it is connected to passion and feeling yet is also able to see like the eye and think like the brain. We never know exactly which of the heart's many manifestations we are dealing with; with the reading of each sonnet, our assumptions must be initially suspended and potentially adapted anew.

In sonnet 46, the jury of the heart's thoughts fails to award it sovereignty over the whole: the "verdict" apportions "The clear eye's moiety and the dear heart's part" more equitably than that. The final resolution, in which eye and heart each get one line of the couplet, seems rather too trite, seems merely to confirm the artificiality of the alleged conflict. At the same time, the terms of the settlement are ambiguous: the eye winds up with the "moiety," and the heart is given a "part." This nebulous separation of "outward part" and "inward love" appears to divide two things that were arguably distinct to begin with. The overblown dichotomous quality of the sonnet has prompted readers to articulate the dialectic that it encodes: for Stephen Booth, it has to do with infatuation and true love; for Helen Vendler, it relates to love's aesthetic and affective aspects.[41] As well, the two disputants could be said to represent different facets of consciousness—perception and contemplation, the eye being a window to the world and the heart the closet of interiority. Sonnet 46's harmonious apportioning suggests that these are not alternatives but rather inextricable aspects of the unified phenomenology of experience.

Perhaps the surest indication that there is no real disagreement in sonnet 46 is that the truce presented in its sequel, sonnet 47, is more compelling than the original quarrel:

Betwixt mine eye and heart a league is took,
And each doth good turns now unto the other.
When that mine eye is famished for a look,
Or heart in love with sighs himself doth smother,
With my love's picture then my eye doth feast
And to the painted banquet bids my heart;
Another time mine eye is my heart's guest,
And in his thoughts of love doth share a part.
So either by thy picture or my love,
Thyself, away, art present still with me,

41. Stephen Booth, ed., *Shakespeare's Sonnets* (New Haven, CT: Yale University Press, 1977), 208–10; Helen Vendler, *The Art of Shakespeare's Sonnets* (Cambridge, MA: Harvard University Press, 1997), 234.

> For thou no farther than my thoughts canst move,
> And I am still with them, and they with thee;
>   Or if they sleep, thy picture in my sight
>   Awakes my heart, to heart's and eye's delight.

Instead of a legalistic dispute, we have reciprocal rituals of hospitality—"famished," "feast," "banquet," "guest," "share." The first two quatrains outline scenarios wherein the organs work to nourish one another. In one, heart and eye feast together on "the painted banquet" of "my love's picture." In the second, the eye shares in the heart's "thoughts of love." How may heart and eye feast jointly on "the painted banquet" of "my love's picture"? It could be that the lover is referring to an actual painting, but we are told in line 13 that the "picture" is obtainable in sleep.[42] He must be speaking of a phantasm, an inward image formed of sense and held in remembrance, which can be passed back and forth between different realms of cognition.[43] The image of the beloved is both visual and intellective; it makes possible a collaboration between "heart" and "eye" that was absent in the preceding sonnet. This more serene portrait of inwardness breaks down the pleasures of recollection into happy undulations, a satisfying cognitive dialectic. In this context, the unexpected emergence of the first-person pronoun near the end—"And I am still with them"—may be signaling a recovered sense of self-possession. The "banquet" is "painted" in more ways than one: it is a picture colored by fond feeling, a rosy phantasm. It is also a literary representation; just as the lover is making mental images, the sonnet renders a verbal picture for us, its readers.

The single-mindedness with which sonnets 46 and 47 pursue the relation between heart and eye can seem overwrought. Booth, for example, finds in their chiastic twists and derivative theme a "barren ingenuity," a "neurotic diversion of energy on to trivia."[44] Joel Fineman says the two sonnets are bound by "a logic of sympathetic opposition": whether the starting point is difference or sameness, the end result is effortless complementarity.[45] What the discourse of imagination reveals is that these contorted interactions between heart and eye have a basis in the equally convoluted characterization of faculty psychology in early modern anatomical culture. Gail Kern Paster

---

42. See Katherine Duncan-Jones, ed., *Shakespeare's Sonnets* (London: Thomas Nelson, 1997), 47, for her interpretation of "my love's picture" as "a concrete depiction," among other meanings.

43. One of the operative meanings of *picture* in the sixteenth century was in fact "mental image." *OED Online*, s.v. "picture, n." 6a. (Oxford: Oxford University Press, January 2018). http://www.oed.com/view/Entry/143501.

44. Booth, *Shakespeare's Sonnets*, 208.

45. Joel Fineman, *Shakespeare's Perjured Eye: The Invention of Poetic Subjectivity in the Sonnets* (Berkeley: University of California Press, 1986), 74.

has written that "that which is bodily or emotional figuration for us . . . was the literal stuff of physiological theory for early modern scriptors of the body."[46] Accordingly, it is only when these poems are read in light of phantasmal theory that they make sense from the perspective of physiological plausibility. Their force emerges when their skirmishes are taken as more than just baroque metaphors and rhetorical displays.

Offering up diverse representations of cognition in which "heart" is sometimes able to stand for "brain" and "mind," just as "eye" can mean "mind's eye," Shakespeare's sonnets iteratively contemplate the ways in which mental representation mediates embodied experience. Sonnet 137, for example, explores the material heft of phantasms:

Thou blind fool love, what dost thou to mine eyes,
That they behold and see not what they see?
They know what beauty is, see where it lies,
Yet what the best is take the worst to be.
If eyes corrupt by over-partial looks
Be anchored in the bay where all men ride,
Why of eyes' falsehood hast thou forgèd hooks
Whereto the judgment of my heart is tied?
Why should my heart think that a several plot
Which my heart knows the wide world's common place?
Or mine eyes, seeing this, say this is not,
To put fair truth upon so foul a face?
        In things right true my heart and eyes have erred,
        And to this false plague are they now transferred.

The opening lines contrast "beholding" and seeing, drawing attention to the sense of touch hidden in *behold*—what exactly is "held" in the eye, or mind's eye? The word creates a "hallucinatory tactility," suggesting the slight texture of the mental image.[47] That slightness contrasts the heavier bonds between perception and understanding: heart's "judgment" is enchained by "forgèd hooks" to "mine eyes." The unusual verb choices imply that these eyes have exceeded their mandate: they not only "see" beauty but seem to "know what beauty is." Capable of knowledge, they nonetheless swap "the best" for "the worst." Either eyes are mistaken in their judgment—they confuse truth and

---

46. Gail Kern Paster, "Nervous Tension," in Hillman and Mazzio, *Body in Parts*, 111.
47. Susan Stewart, *Poetry and the Fate of the Senses* (Chicago: University of Chicago Press, 2002), 163.

lies—or else they correctly detect the beloved's duplicity and choose to look past it. Either way, they have become "corrupt[ed]" in more than one sense. Punitively, the couplet pulls the body to the fore: false perception becomes a "plague," recalling the contemporary belief that the heart could send up harmful spirits to the eyes, transmitting toxic vapors through the gaze.[48] In this sonnet, seeing is not harmless. Even to look on the love object from afar is to be beholden to disease, to become entrammeled in hooks and bonds and weighted with anchors.

In the sensory vacuum of darkness, phantasms detach from the guiding influence of the body altogether, wandering freely. Sonnet 27 focuses on the gauzy play of mental images that are released before sleep:

> Weary with toil, I haste me to my bed,
> The dear repose for limbs with travel tired,
> But then begins a journey in my head
> To work my mind, when body's work's expired.
> For then my thoughts (from far where I abide)
> Intend a zealous pilgrimage to thee,
> And keep my drooping eyelids open wide,
> Looking on darkness which the blind do see;
> Save that my soul's imaginary sight
> Presents thy shadow to my sightless view,
> Which like a jewel hung in ghastly night
> Makes black night beauteous and her old face new.
> > Lo thus, by day my limbs, by night my mind,
> > For thee, and for myself, no quiet find.

Moving from limbs to head, and then to mind, thoughts, and soul, we gradually leave the weary body behind and glide into a bodiless nocturnal "journey." As much as "behold" in sonnet 137, "intend" seems the key choice here—*tend in*. The word is all the more apt because in early modern usage it connotes both fixedness and extension and so evokes perfectly imagination's steady grasping.[49] Even as the lover succumbs to "dear repose," his intending mind unfurls beyond "where I abide"; phantasms do not just swim throughout his corporeal form, they stretch out into the world. A series of paradoxes creates a feeling of otherworldliness: "looking on darkness"; "the blind do see"; "sight-

48. Robert Burton, *The Anatomy of Melancholy*, ed. Thomas C. Faulkner, Nicolas K. Kiessling, and Rhonda L. Blair (Oxford: Clarendon, 1989), 3:88.

49. *OED Online*, s.v. "intend, v." 1 and 18. (Oxford: Oxford University Press, January 2015). http://www.oed.com/view/Entry/97442.

less view." The supernatural and spiritual overtones—"zealous," "pilgrimage," "ghastly"—confirm that, in the deepest recesses of the mind, the visual becomes the visionary. Here, the speaker is captivated by the ability of his "soul's imaginary sight" to project pictures before "eyelids open wide," to hang shadow images in darkness. The "old" face of "black night" grows "beauteous" as the beloved's face fades in and out of it.

The separation of eye and mind's eye is not always so exalted, however. Sonnet 113 illustrates the dichotomy of anatomical function and structure that we saw earlier:

Since I left you, mine eye is in my mind,
And that which governs me to go about
Doth part his function, and is partly blind,
Seems seeing, but effectually is out.
For it no form delivers to the heart
Of bird, of flower, or shape which it doth latch;
Of his quick objects hath the mind no part,
Nor his own vision holds what it doth catch.
For if it see the rud'st or gentlest sight—
The most sweet favor or deformed'st creature,
The mountain or the sea, the day or night,
The crow or dove—it shapes them to your feature.
　　Incapable of more, replete with you,
　　My most true mind thus makes mine eye untrue.

This is the only time in the sonnets that Shakespeare employs the word *function*, a recent coinage in sixteenth-century English and a word that he appears to have associated with sensory incapacitation, writing in *A Midsummer Night's Dream* of "dark night, that from the eye his function takes" (3.2.177).[50] Various ruptures are noted in the opening quatrain—between "I" and "you," between "eye" and "mind," and within the eye's parted "function." Though the speaker is physically intact, all is not as it should be. He can discern a bird, flower, or mountain; however, his mind cannot grasp these "quick objects" for long. Just as quickly, they are replaced by the image of the beloved. He can still see, can still create mental images, but the "forms" apprehended by his eyes are no longer relayed to his "heart." This can occur only if the body's function and form are indeed able to part ways, such that the eye "seems seeing,

50. *OED Online*, s.v. "function, v." 2a. (Oxford: Oxford University Press, January 2015). http://www.oed.com/view/Entry/75476.

but effectually is out." The difference between structure and operation—the gap that Vesalius refuses to close with conjecture and that Du Laurens and Huarte were at pains to resolve—is the source of this poem's ingenuity. The dark lady sonnets often invert opposites such as "fair" and "black"—but here in sonnet 113, day is not exactly mistaken for night, nor crow for dove. The categories have not been switched; they simply do not exist. As we will see in a later chapter, a diseased imagination was historically used to account for precisely this kind of cognitive failure. Medical treatises proposed that, in cases of extreme love melancholy, the phantasmatic love object could become permanently lodged in the mind, crowding out all else and eventually driving the subject mad.[51] This sonnet suggests that such distraction can be self-induced, even pleasurable: having described his sensory breakdown over the course of three quatrains, in the couplet the lover easily exonerates his "mind," calling it "most true."

In sonnet 114, the relation between eye and mind is more overtly founded on deception:

> Or whether doth my mind, being crowned with you,
> Drink up the monarch's plague, this flattery?
> Or whether shall I say mine eye saith true,
> And that your love taught it this alchemy,
> To make of monsters and things indigest
> Such cherubins as your sweet self resemble,
> Creating every bad a perfect best
> As fast as objects to his beams assemble?
> Oh, 'tis the first! 'Tis flatt'ry in my seeing,
> And my great mind most kingly drinks it up.
> Mine eye well knows what with his gust is 'greeing
> And to his palate doth prepare the cup.
>   If it be poisoned, 'tis the lesser sin
>   That mine eye loves it and doth first begin.

The assonance of "mine," "mind," "eye," and "I" ties all these words together in a way that reflects the subverted power structure that the poem delineates. Who is really in control? Ostensibly, the "mind" is king, being "crowned with you." But the monarch could well be the victim of the eye's meddling "flatt'ry." Perception offers a "cup" of tranquilizing error that turns "monsters" into an-

---

51. See Angus Gowland, *The Worlds of Renaissance Melancholy: Robert Burton in Context* (Cambridge: Cambridge University Press, 2006), 66–67.

gels, "bad" into sweet. As we discover, the highest authority in this scene of intrigue is not the "kingly" mind but rather the "I"—the persona that is posing the questions of the sonnet, who discerns the various deceptions in play and observes almost dispassionately the dramatic action between prince and courtier as though in a mental *theatrum*. Cleverly, the poem toys with the political language often seen in early modern psychological discourse. Du Laurens calls the brain "that magnificent and stately turret" that houses "the soueraigne power of the soule," to which all the other bodily parts rightly "pay tribute."[52] What Shakespeare gives us in sonnet 114 is not so much a well-tuned state as a scene of political subversion. He demonstrates an awareness, maybe a wariness, of the ideological motives that underpin theories of cognition.

PROMETHEAN DREAMS

Early modern anatomy certainly had such motives. As Crooke would write, the student of the human fabric stands to enjoy the sanative benefits of self-control and social adjustment. He will be able to "moderate and order the conditions and affections of his minde." Such moderation is "a very glorious thing," though it is also "very hard and difficult" to attain. "And yet," adds the author, "by the dissection of the body, and by Anatomy, wee shall easily attaine vnto this knowledge."[53] These lofty humanistic goals were synopsized in the dictum *Nosce te ipsum*, "Know thyself," to which, as Michael C. Schoenfeldt writes, "much of the physiological work of the period arises as an explicit response."[54] The higher purpose of anatomy was to anatomize the self, as Jonathan Sawday describes: "To open the body of another was, it was true, part of the process of achieving generalized understanding of the human frame and its creator's wisdom, but the words ["know thyself"] also led the enquiring human subject to a form of self-analysis."[55]

These goals were circumscribed by how anatomy was carried out in practice, though. For one thing, in the new anatomical age, introspection and other sorts of conjectural thinking did not count as legitimate means of discovery. Vesalius was especially critical of the fabrications and handed-down myths of his precursors: "Who, I beg you, after diligently examining the parts of the human fabric, who is not dependent upon another's writings or pictures (as

---

52. Du Laurens, *Discourse*, 6.

53. Crooke, *Mikrokosmographia*, 12.

54. Michael C. Schoenfeldt, *Bodies and Selves in Early Modern England: Physiology and Inwardness in Spenser, Shakespeare, Herbert, and Milton* (Cambridge: Cambridge University Press, 1999), 12.

55. Jonathan Sawday, *The Body Emblazoned: Dissection and the Human Body in Renaissance Culture* (London: Routledge, 1995), 110.

surely no one should be) will not perceive on this evidence that Galen made many things up, especially in *De usu partium?*"⁵⁶ In the original Latin, Galen's fault is literally that he "imagined" many things—*pleraque imaginatum fuisse*. To choose not to look at the body's glorious fabric is no less than a hubristic denial of the creator's art.

> For who, immortal God, will not be astonished at the crowd of the phi-
> losophers of this age—and, I might add, of theologians—who make such
> fools of themselves by disparaging the divine and supremely admirable
> machine of the human brain in their impious dreams against the Maker
> of the human fabric no differently than if they were Prometheuses? They
> invent the most witless falsehoods about some structure of the brain,
> the very structure which the infinite Maker of things fashioned with
> incredible forethought and so expertly for the functions it needed to
> perform for the body, keeping it far from their eyes and substituting
> their own version so filled with clumsy monstrosities.⁵⁷

Refusing to see (*ab oculis arcentes*) the human form, philosophers and theologians invent one in its place, succumbing to their Promethean "dreams" and
in the process engendering "monstrosities." The failure is twofold: Galen and
others began in the first place with "writings or pictures" rather than observations; then the notions they came up with were given too much credence.
Text and image may be useful mnemonic aids, wrote the fifteenth-century
anatomist Alessandro Benedetti, but they cannot adequately represent things
in themselves: "Those people who, having trusted only in written accounts
(as Plato says) without an inspection of 'things,' are not turning over the impressed 'things themselves' in their minds; they are frequently deceived and
consign opinion rather than truth to their minds."⁵⁸ The author does not rule
out written notes completely, for they do aid in the recollection of things previously seen. Still, representations are no substitute for "frequent attendance
at public dissections."⁵⁹ These are warnings against imagination: mental images, be they the fantasies of philosophers or verbal pictures printed in books,
cannot produce solid anatomical knowledge.

It is ironic, therefore, that the stunning illustrations of Vesalius's *Fabrica*,
those unforgettable drawings in which flayed musclemen pose in theatrical
attitudes of passion and despair amid pastoral landscapes, are vigorously

56. Vesalius, *Fabric of the Human*, 2:1286.
57. Vesalius, 2:1258.
58. Quoted in Cunningham, *Anatomical Renaissance*, 72.
59. Quoted in Cunningham, 72.

fantastical.[60] So too is the frontispiece of the work, with its teeming hall of figures gathered around an open cadaver while the author-anatomist, standing center stage, looks frankly at the viewer. Visually gorgeous, the book is guilty of the unreality its author professes to disdain. Although empirical means are needed to procure knowledge, then, it would seem that affective representational strategies can be used to display the result. The fruits of anatomical study can be conveyed imaginatively, so long as the information has been gathered in the proper manner.[61] In their own way, the *Fabrica*'s illustrations evoke the confessional attitude of self-anatomy: its animated cadavers support the early modern fantasy described by Sawday, which imagines that the anatomical subject in some sense willingly participates in his or her dissection.[62] Yet there remains an unresolved tension—as one illustration especially implies. The picture is of a "skeletal Hamlet," standing in the familiar pose of the melancholiac, with legs crossed and head leaning against one hand, the other resting on a human skull (figure 3).[63] Needless to say, the two skulls—one of which is sentient, the other not—look exactly the same. A nearby inscription reads, *Vivitur ingenio caetera mortis erunt*: "Genius lives, all else is mortal." An alternative tag line could be "This is, and is not, me"—for what the picture inadvertently signals is the irremediable otherness of the dissected corpse that is supposed to be the instrument of self-knowledge. Another illustration, from Juan Valverde de Amusco's *Anatomia*, makes a comparable point (figure 4). It depicts what appears to be a dissection in progress: a man stands over an open cadaver, his fingers sunken deep in its chest cavity. But, in a twist, the "anatomist's" own torso has been cut open too, the sternum pulled aloft to reveal the rib cage by the invisible hand of another anatomist, the illustrator.[64] Despite the visual similarities between demonstrator and

---

60. Such representations are not unique to the *Fabrica*, or exclusively male. See Sawday, *Body Emblazoned*, for more examples—for instance, the impassive women in Adrianus Spigelius's *De Formato Foeto* (1627), whose abdominal flesh is drawn back to reveal their wombs and ovaries (figures 28 and 30 in Sawday), displaying "blank and unknowing stare[s], refusing to acknowledge [their] subservience to science and representation" (Sawday, 103).

61. Glen Harcourt has described how such illustrations were strategically geared toward legitimizing anatomical practice. See "Andreas Vesalius and the Anatomy of Antique Sculpture," *Representations* 17 (1987): 28–61.

62. Sawday, *Body Emblazoned*, 110.

63. Andreas Vesalius, plate 22, "A Delineation from the Side of the Bones of the Human Body Freed from the Rest of the Parts Which They Support, and Placed in Position," in *The Illustrations from the Works of Andreas Vesalius of Brussels*, ed. J. B. de C. M. Saunders and Charles D. O'Malley (Cleveland: World, 1950), 86–87.

64. Juan Valverde de Amusco, *Anatomia del corpo humano* (Rome, 1559), 108. Also see Cunningham, *Anatomical Renaissance*, 161–62, which notes that the image is a composite of two images from Vesalius's *Fabrica*. I am indebted to Sachiko Kusukawa for pointing this out.

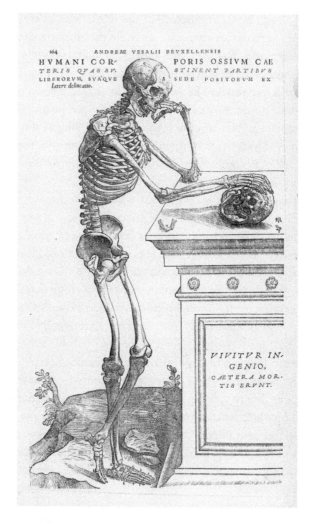

FIGURE 3.   Anatomical illustration. From Andreas Vesalius, *De humani corporis fabrica* (Basel, Switzerland, 1543), 164. Wellcome Library, London.

specimen, the two are alienated from one another: the latter's eyes are blissfully shut, the former's directed heavenward in an abject expression. Such images hint at the complicated ways that anatomy and dissection were shaping Renaissance ideas of subjectivity. Their unspoken ironies resonate with the historical fact that, though the pedagogical ideal for anatomical study was the healthy physique of a well-developed young man, the actual bodies used were typically "executed criminals, paupers, or friendless

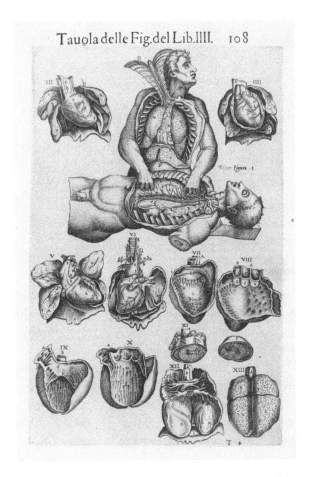

Tauola delle Fig.del Lib.llll.   108

FIGURE 4. Anatomical illustration. From Juan Valverde de Amusco, *Anatomia del corpo humano* (Rome, 1559), 108. RB 498768, Huntington Library, San Marino, California.

foreigners."[65] The ideal of self-analysis is unrealizable, as impossible as grasping the mind's nature by looking at brain tissue; the anatomical body is always an imperfect approximation of the observer, his double and his foil.

The visual politics of autopsy revitalizes the well-worn trope of lovers gazing at one another in Shakespeare's sonnet 24. The poem emphasizes the gulf between subject and object that simultaneously contracts and expands in the moment of observation.

---

65. Nancy G. Siraisi, *The Clock and the Mirror: Girolamo Cardano and Renaissance Medicine* (Princeton, NJ: Princeton University Press, 1997), 96.

Mine eye hath played the painter and hath steeled
Thy beauty's form in table of my heart;
My body is the frame wherein 'tis held,
And perspective it is the best painter's art.
For through the painter must you see his skill
To find where your true image pictured lies,
Which in my bosom's shop is hanging still,
That hath his windows glazèd with thine eyes.
Now see what good turns eyes for eyes have done:
Mine eyes have drawn thy shape, and thine for me
Are windows to my breast, wherethrough the sun
Delights to peep, to gaze therein on thee.
    Yet eyes this cunning want to grace their art:
    They draw but what they see, know not the heart.

As Rayna Kalas points out, this sonnet begins with an anatomical "frame" ("table of my heart") that turns into a two-dimensional perspectival frame; it brings scientific observation together with art.[66] The situation of the sonnet is that the lover has invited the beloved to look inside him—"through the painter you must see." But the view is triply skewed. For one thing, the "image" is hidden inside the workshop of the lover's "bosom." To see it the beloved's gaze must penetrate the lover's body—*perspective*, Vendler points out, denotes a seeing through (*per-spicio*).[67] Second, the inward image is a representation; it might even be a distorted anamorphic "perspective," a puzzle portrait that only resolves into an intelligible picture when viewed from a particular angle. And there is a third problem: the painter's eyes are windows as well as mirrors "glazèd" with the beloved's reflection. Shakespeare plays on the fact that glass is both transparent and reflective, allowing two images to be "seen" in it at once. As enthralling as this superimposition of images might initially seem, it is superficial and illusory. Nothing in this sonnet can be plainly seen or shown: the beloved observes only what the lover's painterly "skill" reveals, the desiring painter-lover's eyes "draw but what they see," not knowing the beloved's true feelings. Empirical observation is

---

66. Rayna Kalas, *Frame, Glass, Verse: The Technology of Poetic Invention in the English Renaissance* (Ithaca, NY: Cornell University Press, 2007), 178. Compare with Constable's *Diana 5*, which employs a similar conceit but with less evocative language:
    Thine eye the glasse where I behold my hart,
    mine eye the window, through the which thine eye
    may see my hart. (C1r)
67. Vendler, *Shakespeare's Sonnets*, 142.

here mixed up with "cunning" and "play"; the "grace" of true understanding is absent. The gaze remains at a remove from what lies beneath, the complex interior designated in the couplet simply as "heart." For Kalas, this sonnet's ocular windows mimic the "reflective transparency" of poesy; for Fineman, the poem epitomizes the idealizing visuality of the Elizabethan lyric.[68] This sonnet is also about the limits and necessary distortions of representation, which obscures even as it attempts to reveal, just as the anatomist's shadow eclipses his object. Characteristically, Shakespeare's speaker seems aware of and even bemused by this—"See what good turns eyes for eyes have done."

The knowledge that the anatomical body is an imagined body resides uneasily in the early modern discourses devoted to its study; yet this is the knowledge to which Shakespeare's sonnets return again and again. Their exploration of mental imagery confirms that empirical discoveries do not diminish but rather redouble our need for satisfying stories about our insides. In their intense programmatic examinations of the ties between body and mind, Shakespeare's sonnets constitute their own sort of anatomical illustrations. With systematic ingenuity, they seek to pinpoint the mutable connections between thought and sense, idea and feeling. At times, the lover shows the anatomist's single-minded focus, being less concerned with noble symmetries of syncretized knowledge than with the discernible realities of his embodied state; at other moments, he comes closer to the skeletal tragedians of the *Fabrica*, embracing performativity and ostentatious display. But the lover is also like the psychologists of Shakespeare's time, committed to forging figurative hypotheses, testing narratives that are intuitively meaningful if not provable. In their investigations of mental imagery, the sonnets belong very much to the period that was the dawn of modern anatomy; just as well, they demonstrate that the paradigm of faculty psychology remained relevant to the Renaissance project of dissecting subjectivity. Shakespeare's lover knows that fancy is not to be found in either heart or eye, because fancy represents a process of thinking about—thinking through—his desiring and disoriented body.

---

68. Kalas, *Frame, Glass, Verse*, 185; Fineman, *Shakespeare's Perjured Eye*, 139.

CHAPTER 2

Children of Fancy

Academic Idleness and *Love's Labor's Lost*

PLAYING WITH TOYS

Often in early modern writings imagination is described as "idle." Take John Florio's 1603 translation of Michel de Montaigne's essay "Of Idlenesse," which observes that minds unburdened by purpose are liable to run wild: "Except they be busied about some subject, that may bridle and keepe them vnder, they will here and there wildely scatter themselves through the vaste field of imaginations."[1] The author can testify to this because his own mind has recently been unbridled, as it were, let free to roam the "vaste field" of fancy: "I retired my selfe unto mine owne house, with full purpose, as much as lay in me, not to trouble my selfe with any businesse, but solitarily, and quietly to weare out the remainder of my well-nigh-spent life; where me thought I could do my spirite no greater favour, than to give him the full scope of idlenesse, and entertaine him as he best pleased." Hoping to have set aside "trouble" and "businesse," Montaigne finds however that his liberated "spirite" is like a "skittish and loose-broken jade," begetting "extravagant *Chimer-*

---

1. Michel de Montaigne, "Of Idleness," in *The essayes*, trans. John Florio (London, 1603), 14. Samson Lennard's translation of Pierre Charron's *Of wisdome* (London, 1608), 56–57, contains a passage uncannily reminiscent of Florio's Montaigne: "The *Spirit*, if it be not busied about some certaine obiect, it runnes riot into a world of imaginations, and there is no folly nor vanity that it produceth not; and it haue not a setled limit, it wandreth and loseth it selfe" ("Of the humane Spirit, the parts, functions, qualities, reason, inuention, veritie thereof").

aes, and fantasticall monsters" that seem "orderlesse, and without any reason."[2] What was intended to be a mental respite devolves into "wilde & vnprofitable" agitation, giving way to "follie, or extravagant raving," the fretful dreams of sick men who "faine, / Imaginations vaine."[3] In characteristic form, Montaigne resolves to make a study of these disruptions, to record introspectively the vagaries of his inward jade. "At-leasure to view the foolishnesse and monstrous strangenesse of them," he says, "I have begunne to keepe a register of them."[4]

In a small space, Montaigne gathers ideas that were often clustered in contemporary discussions of imagination: recreation, knowledge, and maturity. Strictly speaking, his description of a mind put out to imaginative pasture is an exception that proves the rule: more often in Renaissance literature it is the abandonment of, not the return to, fancy that is noted, and this is done at the end of youth rather than in senescence. Typically, toying with fancy is seen as an activity of children, appropriate to the earlier stages of life. As we will see in other texts, the notional idleness of imagination encodes unresolved questions about the faculty's role in the creation of knowledge, which arose as part of the period's epistemological reassessment of different forms and methods of learning. Against this backdrop, Shakespeare conceives a comedy about an abortive "academe," filled not with scholars but with fantastical students, crusty pedants, foppish courtiers, and children. Framed by the rise of early modern empiricism as well as the changing social and institutional topography of Renaissance academia, *Love's Labor's Lost* depicts a school of fancy. The comedy showcases imagination's idiosyncratic method, weighs its power to innovate, and suggests that its role in the derivation of scientific knowledge from sensed realities of the world be refined and remade.

Before turning to the play, it is worth examining the cultural trope on which Shakespeare's satire draws: the idea, namely, that imagination is a leisurely and trifling kind of mental pursuit, which ought to be renounced as a precondition for wisdom and erudition. This idea, which is conventional in sixteenth-century essays and verse, is appropriated by *Love's Labor's Lost* and there translated into a series of observations about the uncertain instructional value of imagination. Perhaps the clearest indication of fancy's associations with puerility is the word *toy*, which denotes an abstract rather than a concrete plaything—recall that "fairy toys" are among the synonyms Theseus offers in *A Midsummer Night's Dream* for the imaginative figment. The image-making

---

2. Montaigne, 15.

3. Montaigne, 14.

4. Montaigne, 15.

faculty is prone to purposeless self-indulgence; it is, as the poet John Davies of Hereford writes, a "whelpe" that "plays with every toy."[5] Like *fancy*, *toy* carries an erotic connotation too. In Shakespeare's *Venus and Adonis*, for instance, the goddess boasts of getting her warlike husband, Mars, "to toy, to wanton, dally, smile and jest."[6] Likewise, newlywed Othello resolves not to be distracted by the "light-winged toys / Of feathered Cupid," the pleasures of married life (1.3.265–66). As we saw in the previous chapter, these amorous and cognitive valences are not mutually exclusive; rather, they often go together. Thus, in John Lyly's *Endymion*, a jilted lover conflates love melancholy with mental idling: "I have no playfellow but fancy, . . . and make my thoughts my friends." Likewise, Thomas Watson's *Tears of fancie* laments that "cruell loue" produces "Idle toyes that tosse my brayne."[7]

The idea of toying is explored in Nicholas Breton's 1577 collection of short poems, *A floorish vpon fancie*. One lyric, entitled "The Toyes of an Idle Head," ponders the folly of fancy in a manner typical of the volume:

> Now by my troth, I cannot chuse but smyle,
> To see the foolish fyttes of Fantasye:
> With what deceites she doth the mynde beguyle,
> As pleaseth best her great inconstancye.[8]

This opening stanza makes some oft-heard complaints: that fancy is foolish, deceitful, and inconstant. With its tone of mock indignation, however, the lyric is also wryly self-conscious. Multiple poems in the collection are entitled "toys," crystallizing the notion that a poem is itself a toy, a trifle of the mind. A favorite trope is the "farewell to fancy," the wistful renunciation of erstwhile fond feeling. Breton has a poem by this very name, which opens like this:

> Fonde *Fancie* now farwell, thy *Lodginge* likes me not,
> I serued thee long full like a slaue, yet litle gaines I got.[9]

---

5. John Davies of Hereford, *Mirum in modum* (London, 1602), D3r.

6. William Shakespeare, *Venus and Adonis*, 106. Shakespeare references are from *The Norton Shakespeare*, ed. Stephen Greenblatt et al. (New York: W. W. Norton, 2016), hereafter cited parenthetically.

7. John Lyly, *Endymion*, in *English Renaissance Drama*, ed. David Bevington, Lars Engle, Katharine Eisaman Maus, and Eric Rasmussen (New York: W. W. Norton, 2008), 4.1.33–35; Thomas Watson, *The tears of fancie* (London, 1593), sonnet 35, C3v.

8. Nicholas Breton, *A floorish vpon fancie* (London, 1577), F2r–F2v. Consider also the full title of Robert Tofte's sonnet cycle *Laura: The Toyes of a traueller* (London, 1597).

9. Breton, *Floorish vpon fancie*, C3r.

The final lines are equally performative, a valedictory speech act: "I now forsake thee quite, . . . and so Fancie, farewell." The grandiosity of the leavetaking gesture is a little undercut, though, by the fact that it has to be repeated several times throughout the book: fancy, it seems, is not easily bidden goodbye.[10] Playfully self-deprecating and disingenuous, Breton's poems are really celebrating what they appear to renounce.

Other texts indicate that the "farewell to fancy" was a more or less recognizable literary device. A somewhat more serious version appears in Anthony Munday's *Banquet of daintie conceits*. Here, the speaker's reason for abandoning fancy is not so much frustration as a sense of resignation about time's onward march:

> Farewell sweet Fancie,
> Thou maist goe play thee,
> Wisedome saith, I may not stay thee:
> I am vnskilfull,
> And thou too wilfull,
> And Time dooth thy sports denay me.
> Olde men haue learned:
> And I my selfe haue this discerned.
> That Sports and pleasure:
> Must be applyed to time and measure.[11]

Life is finite in "measure," the speaker realizes. Although imagination's distractions seem potentially unending, he decides to heed the sage advice of "olde men." Meanwhile, Fancy, personified as a playmate, continues to frolic even as the speaker draws away from "Sports and pleasure," succumbing to "Wisedome" and "Time." Perhaps the best-known valediction to imagination, however, is John Milton's "Il Penseroso." The poem, a meditation on meditation, begins with a necessary tidying of the mind:

> Hence vain deluding joyes,
>   The brood of folly without father bred,
> How little you bested,
>   Or fill the fixed mind with all your toyes;

---

10. The word *farewell* appears twelve times in the volume.

11. Anthony Munday, "A Dittie, Wherein the Author Giueth his Farewell to Fancie," in *A banquet of daintie conceits* (London, 1588), I1r–I1v.

Dwell in som idle brain,
     And fancies fond with gaudy shapes possess.[12]

Like the previous examples, this farewell to fancy includes a ritualistic performative act ("Hence vain deluding joyes"), alludes to children ("brood of folly"), insinuates the mind's foolishness ("fancies fond"), and disapproves of idling. Unlike
Breton, Munday, and Montaigne, though, Milton configures these elements into
a rite of specifically intellectual passage, for it is only by warding off imagination's "toyes" that this contemplative speaker thinks he can attain the "old experience" and "Prophetic strain" to which he aspires.

However difficult it may be to say good-bye, not to do so is ill advised. So
says the English essayist William Cornwallis in his discussion "Of Fantasticknesse": "Fantasticknesse is the Habiliment of youth, Wisdomes minority, Experiences Introduction, the Childe of Inconstancy, the Mother of Attire, of
Behauiour, of Speach spoken against the Haire, Customes Enemy, It is Greene
Thoughtes in Greene years."[13] With a list of epithets, the author equates
imagination with the incipient, the undeveloped, and the immature: "youth,"
"minority," "introduction," "Childe," "Greene." Eager to play, fantasy makes
the body its doll: "as Children doe with Babies," so the image-making faculty
"puts on and off, dresses, and vndresses, layes it to sleepe, and takes it vp
againe all at an instant."[14] Imagination's inclination for toying does not diminish with age, either: even "in spight of wisdom, and gray haires, it will
daunce at three score years olde, and weare Greene, and play with a Feather."
Cornwallis scorns the man who remains fancy's fool in his dotage, calling him
"monstrous, and ridiculous," grotesquely out of step with his life: "I can remember no sight more offensiue to me then a variable old man, that can
speake of nothing but the fashions of his Time, the wench then in price, how
many hacks he hath had in his Buckler in a Fleete-streete fray, or the friskes of
the Italian Tumblers."[15] Not only is this "variable old man" caught up in
"fashions" and "friskes," the fashions in question are the stale delights of yester-
day. Lurking in Cornwallis's warning is the possibility of mental disorder. As
Robert Burton would later note in *The Anatomy of Melancholy*, "delightsome . . .
toies" are addictive. They can keep the mind "whole daies and nights without
sleepe, even whole yeares alone in such contemplations, and phantasticall
meditations," gradually impeding "ordinary taskes and necessary businesse"

---

12. John Milton, *Complete Shorter Poems*, ed. Stella P. Revard (Malden, MA: Wiley-Blackwell, 2009), 53.

13. William Cornwallis, *Essayes* (London, 1600), N1r.

14. Cornwallis, N1v.

15. Cornwallis, N4r.

until "any study or employment" is no longer possible.[16] This is the obsessional quality to which Paulina refers in *The Winter's Tale* when she castigates jealous Leontes for indulging "fancies too weak for boys, too green and idle / For girls of nine" (3.2.178–9). Cornwallis concludes, therefore, by advising that the "Childe of Fancy" would do best to give imagination up at the onset of puberty: "In a word, after twice shauing, at the third, Fantasticknesse is to be abandoned, for it is Time to put the wit to Schole, and to leaue playing with these vndigested Apes of the Fancie, to trust to Vertue, not to a French Doublet: If we do thus, it is no harme to haue beene once otherwise."[17] We end, again, with the happy renunciation of toys. The author's benevolence— "it is no harme to haue beene once otherwise"—is perhaps a little too easy, too confident in the belief that fancy's eccentricity will be neutralized eventually. In the conflation of fancy and infancy lies the sanguine expectation that both will be overcome as a matter of course.

According to the foregoing examples, imaginative dallying is opposed to learning and schooling, productive work, and intellectual growth. This connects with the intuitive assumption that knowledge is the product of labor, an intuition reified in the way the early moderns understood the disciplines. For example, Pierre de la Primaudaye's *French academie* praises the diligent scholars of antiquity in a chapter entitled "Of Sciences, of the studie of Letters, and of Histories." The classical thinkers, he says, understood the "greatnes & difficultie of knowledge" and therefore expended "great paine and trauell, that their labor might become profitable" for their successors. It is these men, adds La Primaudaye, men who "spent their life euen with sweating, in seeking out the secrets of nature, and were desirous to ease mans studie," who "diuided science for vs into diuers parts." The shape of early modern learning is, in other words, the fruit of the hardworking ancients. As their inheritors, "we must shunne that idlenes and rechlesnes which is in many, who by reason of the difficultie, which they heare say is in sciences, . . . remaine as buried in ignorance."[18] According to this logic, idleness and imagination stand to upend the labor by which the prior division of the sciences was wrought.

Of course, the order of the disciplines would be upturned during the Renaissance; the "sciences," by which La Primaudaye means "Philosophie, Rhetoricke and Mathematicke," would be jostled by new methods and areas of study. Instrumental to the early modern reconsideration of the arts and sciences was Francis Bacon, who thought at length about the relation between

16. Robert Burton, *The Anatomy of Melancholy*, ed. Thomas C. Faulkner, Nicolas K. Kiessling, and Rhonda L. Blair (Oxford: Clarendon, 1989), 1:243.

17. Cornwallis, *Essayes*, N4v. "Childe of Fancy" appears on M1r.

18. Pierre de la Primaudaye, *The French academie*, trans. Thomas Bowes (London, 1586), 75–77.

university learning and scientific knowledge, as well as the place of imagination in both. In *The Advancement of Learning*, for example, Bacon memorably posits that the faculties of imagination, reason, and memory are respectively responsible for poetry, philosophy, and history.[19] His view of "poesie" is not ungenerous: notably, he elevates poetry above history because fiction is able to contrive moral lessons in a way that chronicles cannot.[20] Still, Bacon's view of imagination is primarily oriented toward the creation of epistemic knowledge, and he was not able to find "any science that doth properly or fitly pertain to the Imagination." Even poesy is "rather a pleasure or play of imagination, than a work or duty thereof."[21]

Not unlike Bacon, Shakespeare in *Love's Labor's Lost* asks what can be learned by way of the image-making faculty. Assembling at the outset an "academe" ostensibly concerned with the cultivation of knowledge, the play proceeds to portray scenes of imaginative toying rather than studious application. In Armado's modishness and Holofernes's bookishness, in Mote's precocity and the lovers' erotic flyting, Shakespeare delineates elements of fancy's distinctive style and describes the hallmarks of the imaginative method, so to speak: its chaotic caprices, its resistance to structure, its flirtation with novelty, and its perennial risks of fruitlessness. Importantly, the play situates fancy in the concrete world of social relations rather than among abstract pedagogical and philosophical ideas, reflecting the changing landscape of Renaissance knowledge making. The hyperactive fantasy, normally wisdom's foil, has a pervasive though implicit presence in Shakespeare's comedy, which attempts, laboriously yet playfully, to articulate imagination's covert role in the pursuit of *scientia*.

A LITTLE ACADEME

We begin with a martial call to fame. The war in question, says King Ferdinand to his comrades in arms, is "against your affections," the "huge army of the world's desires."

> Navarre shall be the wonder of the world;
> Our court shall be a little academe,
> Still and contemplative in living art. (1.1.12–14)

---

19. Francis Bacon, *The Advancement of Learning*, in *The Works of Francis Bacon*, ed. James Spedding, Robert Leslie Ellis, and Douglas Denon Heath, vol. 3 (London: Longman, 1857–1874), 329.

20. Sachiko Kusukawa, "Bacon's Classification of Knowledge," in *The Cambridge Companion to Bacon*, ed. Markku Peltonen (Cambridge: Cambridge University Press, 1996), 54.

21. Bacon, *Advancement of Learning*, 4:382.

Through a program of corporeal mortification—"The mind shall banquet, though the body pine" (1.1.25)—these "brave conquerors" will be transformed into an intellectual and aesthetic "wonder of the world," a *tableau vivant*, "still and contemplative in living art." Lofty idealism aside, the specific aims of the King's "edict" are not clear: we are not told what these scholars will do, only what they will have to give up in order to do it. Biron, who is reluctant to sacrifice food, sleep, and the company of women, immediately objects to the proposed regimen. In his reluctance to swear the oath of austerity, he expresses an epistemological qualm:

BIRON. What is the end of study? Let me know.
KING. Why, that to know which else we should not know.
BIRON. Things hid and barred, you mean, from common sense? (1.1.55–57)

Biron's passing invocation of faculty psychology ("common sense") suggests that knowledge has to be extracted from the mind's more immediate and oftentimes wrong impressions. This knowledge is "hid and barred" from the *sensus communis* in that it has to be obtained by the conscious work of the higher faculties of understanding, memory, and imagination.

Comically, the two men are talking at cross purposes. When Ferdinand speaks of that "which else we should not know," he means to say—as many thinkers of the Renaissance believed—that the highest knowledge is, at least at the outset, concealed from the human intellect. Biron, however, is referring to the kind of knowledge that is held at bay for the sake of social decorum—the sensual knowledge of the libertine:

Come on, then. I will swear to study so,
To know the thing I am forbid to know,
As thus: to study where I well may dine,
When I to feast expressly am forbid;
Or study where to meet some mistress fine,
When mistresses from common sense are hid;
Or, having sworn too hard-a-keeping oath,
Study to break it and not break my troth.
If study's gain be thus, and this be so,
Study knows that which yet it doth not know. (1.1.59–68)

Wanting to "know the thing I am forbid to know," Biron is coyly bargaining for hedonistic license, hoping to loosen some of the puritanical strictures of the contract he is expected to sign. Indirectly, though, he points out that real knowledge is "hidden," thwarted as much by adopted notions of propriety as it is by the mind's native intellective limitations. Also buried in Biron's niggling is a frustration that will be familiar to the especially curious student: within the constraints of institutionalized learning, edification is always restricted by the already settled question of what knowledge is, is automatically stifled by the fact that the very meaning of knowledge has been determined in advance by somebody other than the student. Study already "knows that which yet it doth not know"; it typically reinforces its own paradigmatic prejudices. There is cynicism here as to whether the discovery of new things is even possible. Biron then draws a distinction between received knowledge and knowledge garnered through experience:

> Small have continual plodders ever won,
> Save base authority from others' books.
> These earthly godfathers of heaven's lights,
> That give a name to every fixèd star,
> Have no more profit of their shining nights
> Than those that walk and wot not what they are.
> Too much to know is to know nought but fame,
> And every godfather can give a name. (1.1.86–93)

Who needs learning in order to admire the night sky? The "shining" stars are not diminished by our not knowing their names. Conversely, the dubious "fame" garnered by the "earthly godfathers" who pronounce these names— as well as the less illustrious "plodders" who glean them from "others' books"— is little more than a "base authority."

The play's opening scene thus gestures at salient questions that came to the fore in late sixteenth-century epistemology. By what methods is knowledge created, and what are the limitations of those methods? Where is learning produced, and by whom? As though to underline the significance of these matters, Shakespeare dubs the King's enterprise an "academe"—a word that he appears to have coined—and situates the play's action in a royal court.[22] These decisions are crucial, for they index shifting attitudes toward knowledge production in early modern Europe. It was in this period that science went from being a "kind of knowledge," which one possesses, to a "form of activity," which one carries out.[23] This in turn spurred the need for novel locales in

---

22. *OED Online*, s.v. "Academe, n." 1a (Oxford: Oxford University Press, January 2018). http://www.oed.com/view/Entry/875.

23. Owen Hannaway, "Laboratory Design and the Aim of Science: Andreas Libavius versus Tycho Brahe," *Isis* 77, no. 4 (1986): 586.

which natural phenomena could be investigated. It would not be until the end of the seventeenth century that the laboratory—a term derived from the Latin word for work—would emerge as a premier site of discovery. Before then, protoscientific work was being done in a range of other places, from "the apothecary's and instrument maker's shop, to the coffeehouse, the royal palace, the rooms of college fellows, and associated collegiate and university structures."[24] Elizabethan London was an active site of empirical culture, as Deborah Harkness has described. Away from the court and the universities while still in a cosmopolitan setting that proffered access to the latest mathematical books, residents of Shakespeare's city "were busy constructing ingenious mechanical devices, testing new medicines, and studying the secrets of nature." The labor, tools, and techniques necessary "to shift the study of nature out of the library and into the laboratory" had historically begun.[25]

The fact that *Love's Labor's Lost* is set against a courtly backdrop heightens the epistemological tensions in a different way: it sharpens the play's satirical examination of fancy's intellectual sidelining. In Italy, court patronage fostered scientific experimentation; among the noted beneficiaries of this system were Galileo Galilei and Ulisse Aldovrandi, who spent time at the Medici court, as well as Tycho Brahe and Johannes Kepler, who were attached to the court of the Holy Roman emperor. The early modern period also saw the rise of courtly academies in Italy—among them, the Accademia Secretorum Naturae (established in 1560) and the Accademia dei Lincei (1603). As historians have described, certain conditions of the Renaissance academy made it conducive to innovation: these were communities founded on the values of courtly etiquette; they were also egalitarian "islands of equality" in the midst of otherwise hierarchical societies.[26] The academies enjoyed relative intellectual autonomy away from the universities, which frowned upon novelty and criticism of ancient authorities, and were free to explore "new topics and directions of natural inquiry."[27] Governed by codes of conduct "derived from legal practice, from courtesy manuals, or from epistolary convention," the academies emphasized "civility, conversation, and consensus"; they departed from

---

24. Steven Shapin, "The House of Experiment in Seventeenth-Century England," *Isis* 79, no. 3 (1988): 378.

25. Deborah Harkness, *The Jewel House: Elizabethan London and the Scientific Revolution* (New Haven, CT: Yale University Press, 2007), 2.

26. William Eamon, "Court, Academy, and Printing House: Patronage and Scientific Careers in Late-Renaissance Italy," in *Patronage and Institutions: Science, Technology, and Medicine at the European Court, 1500–1750*, ed. Bruce T. Moran (Rochester, NY: Boydell, 1991), 43.

27. See Bruce T. Moran, "Courts and Academies," in *Cambridge History of Science*, vol. 3, *Early Modern Science*, ed. Katharine Park and Lorraine Daston (Cambridge: Cambridge University Press, 2006), 253; see as well 257, 266, 271.

the more bellicose, increasingly outmoded procedures of scholastic disputation.[28] In imagining a "little academe" at the Navarrese court, therefore, Shakespeare's play engages with the decentralization of contemporary epistemological culture. It announces a preoccupation not only with knowledge creation in general—with the role of experience and observation in the service of invention—but also with the social contexts in which discovery can occur.

If the academies separated themselves from traditional early modern curricula, *Love's Labor's Lost* further distinguishes itself as an unusual sort of academy. This becomes evident if we compare the opening of Shakespeare's comedy with that of La Primaudaye's encyclopedic *French academie*, first published in English in 1586.[29] The latter work is framed as a dialogue among four aristocratic scholars, "being of kin, and neare neighbours, and in a manner of one age," who have been "by the care and prudence of their fathers brought up and nourished together from their yong yeares in the studie of good letters." Eschewing the "excessive libertie" of "the Vniversities of this Realme," these youths steer away from the "ordinarie" and "tedious" learning taught by the "French Colledges"; instead they are privately tutored in "learning and knowledge" and taught "the doctrine of good liuing, following the traces and steps of vertue." In practice, this instruction includes Latin and Greek, as well as "the Morall philosophie of ancient Sages and wisemen, together with the vnderstanding, and searching out of histories." Time is also spent on courtly pursuits, "as namely, to ride horse, to run at the ring, to fight at barriers, to apply themselues to all kinde of weapons, and to follow the chace of beasts." As La Primaudaye's genteel scholars discourse about "how we ought to gouerne our selues wisely and duetifully in all humane actions and affaires," they cover subjects ranging from the body and soul of man (including the mental faculties) to the natural world and cosmos, as well as topics in social deportment and political philosophy suited to the education of princes.[30]

---

28. R. W. Serjeantson, "Proof and Persuasion," in Park and Daston, *Cambridge History of Science*, 3:171. For further discussion of the impact of academic manners in the history of science, see Lorraine Daston, "Baconian Facts, Academic Civility, and the Prehistory of Objectivity," *Annals of Scholarship* 8 (1991): 337–63; and Mario Biagioli, "Etiquette, Interdependence, and Sociability in Seventeenth-Century Science," *Critical Inquiry* 22, no. 2 (1996): 193–238.

29. Several French editions of the work had already appeared before this, beginning in 1577. See Stuart Gillespie, *Shakespeare's Books: A Dictionary of Shakespeare Sources* (New Brunswick, NJ: Athlone, 2001), 277, which notes that La Primaudaye's compendium may have been among Shakespeare's sources.

30. La Primaudaye, *French academie*, 1–4. Bruce R. Smith writes that La Primaudaye was "the closest thing late sixteenth, and early seventeenth-century English readers had to *The Encyclopedia*

Like the academicians sketched by La Primaudaye, Shakespeare's Ferdinand, Biron, Longueville, and Dumaine are aristocrats who embark on an intellectual mission. But the similarities mostly end there. In contrast to their French counterparts, who are characterized by humility and asceticism, the scholars of *Love's Labor's Lost* are inclined to immoderation and overconfidence.[31] In the play, ancient worthies who ought to be moral authorities become the butt of jokes, while "humane actions and affairs"—for example, the politico-economic relations between Navarre and France—remain secondary to the pleasures of wooing. Shakespeare's scholars belong to what Richard Helgerson describes as a generation of Elizabethan "prodigals" who "were trying to reconcile their humanistic education and their often rebellious tastes and aspirations," indulging in "the kind of profitless writing that earned them their celebrity."[32] Shakespeare's academe parts ways with La Primaudaye's utopic ideal and falls short as well of the historical academies of the Renaissance. But what if we were to take Shakespeare's creation for what it is? If this is not an ordinary sort of academy, then what kind of academy is it?

In answer, we might look for a moment at an intriguingly titled lyric in Breton's *Floorish vpon fancie*: "The School of Fancy." As the poet imagines it, Fancy's classroom is filled with lovestruck "Schollers." Though the room is illumined by the fire of "Deepe desire," it is intentionally kept too dark to see clearly. Besides the books that "there are taught," there is "none other sight." Of the students, nearly "all are young," barring the occasional "doting foole." At one point, the speaker of the poem is himself "dandled" on Fancy's knee. With the help of her assistants Folly and Frenzy, Fancy instructs her pupils

to speake, to write, and to indight, to labor and take paine:
To go, to run and ride, to muse, and to deuise,
To iuggle with a déerest freend, to bleare the parents eyes.
To spend both landes and goods, to venter Lim and life,
To make foes frends, and twixt dere frends, to set debate & strife:
To doo and vndoo too, so that they may obtaine
Their mistresse loue: and neuer care, for taking any paine.

Britannica." "Speaking What We Feel about *King Lear*," in *Shakespeare, Memory, and Performance*, ed. Peter Holland (Cambridge: Cambridge University Press, 2006), 26.

31. Rolf Soellner, *Shakespeare's Patterns of Self-Knowledge* (Columbus: Ohio University Press, 1972), 79. See as well John Michael Archer, "*Love's Labour's Lost*," in *A Companion to Shakespeare's Works*, ed. Richard Dutton and Jean E. Howard (Oxford: Blackwell, 2003), 3:321, which suggests that the play counters La Primaudaye's "absurd seriousness" with redolent excesses reminiscent of Plato's *Symposium* and *Phaedrus*.

32. Richard Helgerson, *The Elizabethan Prodigals* (Berkeley: University of California Press, 1976), 3, 11.

To set in braue attire, to please their mistris eye:
Although perhaps they vtterly, vndoe themselues therby.
To learne to sing and daunce, to play on Instruments,
To speake choyce of straunge languages, to try experiments
Straunge, seldome had in vse: in fine, to tell you plaine,
To doo almost they care not what, there ladies loue to gaine.[33]

This educational program partly aligns with the recreational calendar of the Renaissance courtier: there are physical sports and pastimes (riding, singing, dancing), as well as tests of skill (juggling, languages, debate). Yet there are inconsistencies too: some of the slated activities involve feats of invention— to "deuise," "to try experiments"—while others seem counterproductive—"to doo and vndoo," "to set debate & strife." It is in the matter of learning that Fancy's scholars are the most seriously shortchanged. Not only are they trained to be idle and trifle with "toies, / In things of value, little worth," they are misled epistemologically, made "to think blacke white, & wrong for right." In a further irony, their pointless schedule is physically and mentally grueling. The speaker regrets having so much "labor lost" during his arduous tenure at Fancy's academy, but he takes solace in at least being able to warn others against it: "Lose no more labor so, in sutch a witles schoole."[34]

The spirit of Shakespeare's comedy, whose title also mourns a fruitless expenditure of "labor," is closer to that of Breton than that of La Primaudaye. Whereas the courtier-scholars of *The French academie* are young, infantilized by their physically absent but controlling fathers, Shakespeare's students are puerile in their unwillingness to bid fancy farewell. Whereas La Primaudaye's pupils are hard workers, spending "oftentimes the whole night vpon the well studying of that which they purposed to handle,"[35] Shakespeare's group abandons its odyssey of industry before the journey even begins. And whereas the work of *The French academie* proceeds by way of Socratic dialogue, the bookmen of Navarre adopt an assortment of methods, including witty language, poems, and theatrical disguises. Manifest in this play are several elements of Renaissance epistemological culture: the ritual sequestration of juvenile fancy, the academic bias against imaginative play, the emergence of the courtly academy as a new site of intellectual inquiry, an openness to exploring new ways of knowledge gathering. The elements have been rear-

---

33. Breton, *Floorish vpon Fancie*, B1r.
34. Breton, B3v.
35. La Primaudaye, *French academie*, 3.

ranged, however, in a manner that reflects the unclear place of imagination in early modern science.

The case made by Biron against Ferdinand's stern statutes turns out to be convincing. "How well he's read to reason against reading!" marvels the King (1.1.94), who good-humoredly offers to let his friend opt out. Biron refuses this reprieve. In fact, he signs his name right as he learns that the French Princess's embassy is fast approaching—which means, of course, that the men are soon to be forsworn. Notably, then, the charter is ratified only after it is decidedly doomed, after the four lords have agreed that "study" in the form they pledge cannot produce new knowledge. In one sense, the agreement is infelicitous, made in bad faith, following what Patricia Parker has termed a logic of "preposterous reversal."[36] In another sense, the scholars of Navarre have tacitly made a pact to challenge rather than obey epistemological norms, have surreptitiously vowed to flout the reigning ideology. Shakespeare has set up a mock academy where the caprices of fancy will rule and which will showcase the laborious idling of imagination in a way that interrogates its utility in the furthering of knowledge.

FANATICAL PHANTASIMES

One sign that the rules of this academy will be more honored in the breach than in the observance is that the conversation turns to matters of fun even before the contractual ink has dried:

> BIRON. But is there no quick recreation granted?
> KING. Ay, that there is. Our court, you know, is haunted
> With a refinèd traveler of Spain,
> A man in all the world's new fashion planted,
> That hath a mint of phrases in his brain;
> One who the music of his own vain tongue
> Doth ravish like enchanting harmony;
> A man of compliments, whom right and wrong
> Have chose as umpire of their mutiny.
> This child of Fancy, that Armado hight,
> For interim to our studies shall relate
> In high-borne words the worth of many a knight
> From tawny Spain, lost in the world's debate. (1.1.159–71)

---

36. Patricia Parker, *Shakespeare from the Margins: Language, Culture, Context* (Chicago: University of Chicago Press, 1996), 30.

If imagination is a source of recreation for the mind, then it is fitting that these scholars' "interim" entertainments should be provided by a "child of Fancy." Don Adriano de Armado is charged with a task to which Bacon might have felt the imaginative faculty very well suited: "minstrelsy." The Spaniard is going to deliver "music" with his "tongue," telling in "high-borne words the worth of many a knight." Armado's fancifulness is evident in his words, his "mint of phrases"—Biron calls them "fire-new." Soon enough we hear for ourselves his coinages: "festinately," "enfreedoming," "couplement," "infamonize" (3.1.5, 114; 5.1.529, 667). Through these verbal toys, Shakespeare's child of fancy, like the one profiled in Cornwallis's essay, is characterized by novelty and inconstancy; he is a man of "fashion" who wavers blithely between "right and wrong." Comically allegorized in Armado is the dubious value of new rather than old knowledge. In Armado's adversary Holofernes, we will see, bookish wisdom is critiqued too. Excesses of fancy, we find, are located at both of these epistemological extremes.

Novelty and courtliness are tied together in the name by which Armado is called more than once: *phantasime*. This enigmatic term, which would appear to share the etymology of *phantasm*, is not only a neologism but a nonce word—no one seems to have used it besides Shakespeare.[37] The first time, it is spoken by the Princess's attendant, Boyet, who calls Armado

> a Spaniard that keeps here in court:
> A phantasime, a Monarcho, and one that makes sport
> To the Prince and his bookmates. (4.1.96–98)

From context, a "phantasime" is a figure of "sport," a distraction for "bookmates" and enemy of studious plodding. The word is mentioned a second time by Holofernes, the schoolmaster, who complains that Armado "draweth out the thread of his verbosity finer than the staple of his argument. I abhor such fanatical phantasimes, such insociable and point-device companions, such rackers of orthography: as to speak 'dout' *sine* 'b,' when he should say 'doubt'; 'det' when he should pronounce 'debt'" (5.1.16–21). "Insociable" and "point-device" suggest linguistic newfangledness. Both these meanings of *phantasime*—courtly pretension and rhetorical trendsetting—are borne out by *Romeo and Juliet*, in which Mercutio scornfully describes as "fantasies" those "new tuners of accent" and "fashionmongers" who "stand so much on the new form that

---

37. *OED Online*, s.v. "phantasim, n." (Oxford: Oxford University Press, January 2018). http://www.oed.com/view/Entry/142180.

they cannot sit at ease on the old bench" (2.3.26–32).[38] Shakespeare was not the only one to have associated fashion mongering with accent tuning: Montaigne's essay "Of the Institution and Education of Children" equates the impulse to apparel oneself "in some particular and vnusuall fashion" with the desire to "hunt after new phrases, and vnaccustomed-quaint words." Both are "scholasticall and childish" flaws.[39] The two ideas are likewise yoked by Thomas Dekker:

> Fantasticke complement stalkes up and downe,
> Trickt in out-landish Fethers, all his words,
> His lookes, his oathes, are all ridiculous,
> All apish, childish, and Italianate.[40]

For Dekker, "Fantasticke complement" is a figure of courtesy and civility ("complement") who is nonetheless "ridiculous" in looks ("out-landish Fethers") and words ("oathes"). Poor words and looks exacerbate each other, in fact: gentility is marred by juvenile prattle, and attempts at innovative eloquence are undercut by foppish ostentation.

In a similar way, Shakespeare's Armado caricatures the courtly claim to the intellectual vanguard. The period had begun what Anthony Grafton describes as a demotion of classical knowledge, scholasticism, and even Renaissance humanism; these older forms of knowledge would be later dismissed by Bacon and René Descartes as "merely literary," not "the sort of knowledge that gave men power over nature."[41] Appropriately, then, Armado's fantastic fashions are at once integral and marginal to the world of Love's Labor's Lost. He is important in that he is an essential fixture of the curriculum: it should be remembered that Armado, however foolish he may seem, is neither a court jester nor a rustic clown—like the four male protagonists, he too has "promised to study three years with the Duke" (1.2.34). At the same time, the Spaniard is defined by his exclusion; he is an intellectual foreigner as well as an ethnic one. He symbolizes—as per the intention of the four patrons who make him their

38. Q1 has "fantasticoes" rather than "fantasies"—another word apparently coined by Shakespeare: see OED Online, s.v. "fantastico, n." (Oxford: Oxford University Press, January 2018); http://www.oed.com/view/Entry/68118. See also Thomas Dekker's Pleasant comedie of old Fortunatus (London, 1600), E1, which suggests that fantastico is more or less synonymous with phantasime: the eponymous hero says he has "reueld with kings, daunc'd with Queenes, dallied with Ladies, worne straunge attires, seene fantasticoes."

39. Michel de Montaigne, "Of the Institution and Education of Children," in Essayes, 84.

40. Dekker, Old Fortunatus, E2r.

41. Anthony Grafton, Defenders of the Text: The Traditions of Scholarship in an Age of Science, 1450–1800 (Cambridge, MA: Harvard University Press, 1991), 1.

recreational toy—the intended demolition of an unpopular educational program. Shakespeare's liminal phantasime haunts the fringes of the Navarrese academy, just as fantasy lurked at the peripheries of the disciplinary universe, obliquely testing assumptions about the nature and provenance of knowledge.

The university system, meanwhile, is invoked by Armado's antagonist, Holofernes. Clearly wishing to distinguish himself from the dilettantish Spaniard, the schoolmaster huffs at Armado's scholastic pretensions, lumping him with "point-device companions," "rackers of orthography," and other "insociable" types. This last jibe would have been an especially cutting insult to a well-bred courtier. The other two might as easily apply to Holofernes himself, a man who, on the occasion of the King's hunt, devises a tongue twister— "The preyful Princess pierced and pricked a pretty pleasing pricket" (4.2.51)—that he follows up with several more verbal contortions, bending the fallen "pricket" deer into a sorrel deer:

> The dogs did yell; put "l" to sore, then "sorrel" jumps from thicket.
> Or pricket, sore or else sorrel, the people fall a-hooting. (4.2.53–54)

With these tedious puns, it may be that Shakespeare is riffing on the early modern figuration of knowledge seeking as a kind of hunt.[42] Here, *venatio* does not yield secrets of nature so much as offer an occasion for unbridled verbal toying. The chase after ingenuity goes nowhere—Holofernes's interlocutors have difficulty following his meaning, even. Interestingly, Holofernes ties his love of wordplay implicitly to imagination: he says he has "a foolish, extravagant spirit, full of forms, figures, shapes, objects, ideas, apprehensions, motions, revolutions. These are begot in the ventricle of memory, nourished in the womb of *pia mater*, and delivered upon the mellowing of occasion" (4.2.60–65). "Forms, figures, shapes, objects, ideas, apprehensions" are all terms that could denote phantasms; Theseus uses some of them when he speaks of lunatics, lovers, and poets. It sounds, therefore, like the special "gift" Holofernes believes himself to possess is fancy. Armado would certainly appear to think so, for he chides the schoolmaster for being "exceeding fantastical—too, too vain" (5.2.526–27). But Holofernes's understanding of faculty psychology is somewhat peculiar. He says his wit is made in "the ventricle of memory" and "nourished in the womb of *pia mater*"—although, in contemporary accounts, the *pia mater* is more often a membrane, not a cav-

42. William Eamon, *Science and the Secrets of Nature: Books of Secrets in Medieval and Early Modern Culture* (Princeton, NJ: Princeton University Press, 1994), 269.

ity, and fancies are almost never described as being "begot[ten]" in memory.[43] Perhaps there is a statement here about bibliographic knowledge, knowledge traditions that emphasize the authority and importance of books above all: this particular schoolmaster mistakes citations from memory for the creative flourishes of imaginative invention.

If Holofernes has some of Armado's fantastical grandiloquence, then Armado, conversely, has a scholarly streak. The Spaniard complains of being melancholy—conventionally, a scholar's affliction. He trades in Latinisms ("*Veni, vidi, vici*") and enjoys the vigorous questioning of the schoolroom: "Who came? The King. Why did he come? To see. Why did he see? To overcome" (4.1.70–71). In early play texts of *Love's Labor's Lost*, Armado and Holofernes are identified respectively as "Braggart" and "Pedant," stock figures of the *commedia dell'arte*. In a way, these two personas evoke two versions of the early modern man of science, as described by Steven Shapin: the gentleman and the scholar. Whereas the former is sociable, civic-minded, and well mannered, the latter is reclusive, dogmatic, and aloof. Shakespeare's Holofernes fits Shapin's portrait of the pedant: "disputatious, litigious, affected and hectoring"; "temperamentally unbalanced"; "obsessively concerned with matters of little interest to polite society"; "a figure of fun and ridicule"; "a bore"; "a fool, a bookish idiot."[44] The trouble is that some of these attributes belong to Armado as well.[45] In these two figures, Shakespeare embodies the tension in Renaissance literature and culture between what Lucy Munro describes as venerable archaism and grotesque neologism.[46]

This ambivalence also captures the epistemological uncertainty of the time—specifically the growing suspicion that the ancients had been ignorant or wrong on many fronts. On this point, Shakespeare's play anticipates a concern that would be explored more explicitly in seventeenth-century philosophies

43. La Primaudaye, for instance, states that memory's purpose is "to receiue and to hold fast" images created elsewhere in the brain. *The second part of French academie*, trans. Thomas Bowes (London, 1594), 163. On the *pia mater*, see, for example, Christopher Langton, *An introduction into phisycke* (London, 1545), xxiii, which describes the *pia mater* as "an excedynge fyne skynne, made of synowes, . . . compassyng the hole substance of the brayne." See as well Thomas Vicary, *A Profitable treatise of the anatomie of mans body* (London, 1577), D1v, which describes it as a nutritive tissue of blood vessels that "nourisheth the brayne and feedeth it."

44. Steven Shapin, "'A Scholar and a Gentleman': The Problematic Identity of the Scientific Practitioner in Early Modern England," *History of Science* 29 (1991): 290.

45. This is observed by Kristian Smidt, "Shakespeare in Two Minds: Unconformities in *Love's Labour's Lost*," *English Studies* 65, no. 3 (1984): 206–7. Smidt calls the two characters "duplicates," Holofernes being "technically superfluous" to the play, and notices that it is only on one occasion that Shakespeare's so-styled braggart actually brags.

46. Lucy Munro, "Antique/Antic: Archaism, Neologism and the Play of Shakespeare's Words in *Love's Labour's Lost* and *2 Henry IV*," in *Shakespeare's World of Words*, ed. Paul Yachnin (London: Bloomsbury, 2015), 77–101.

of science. That the pedant and the phantasime are more similar than different would perhaps not have surprised Bacon, for example, for they embody two of the three epistemological "diseases" that he diagnoses in *The Advancement of Learning*. One of these is "contentious learning," which is to say, "when men study words, and not matter." Another is "delicate learning," manifested by those "taking liberty to coin, and frame new terms of art to express their own sense."[47] Contentious and delicate learning respectively correspond to the dual threats of "Antiquity" and "Novelty"—"strictness of positions" versus the "strangeness of terms."[48] In hectoring Holofernes and babbling Armado, we have something like the Baconian Scylla and Charybdis of bad science, well before Bacon.

The other disease listed in the *Advancement*, incidentally, is "fantastical learning"—Bacon's name for pseudosciences such as "Astrology, Natural Magic, and Alchemy," which prey on the "vain imaginations" of the credulous.[49] As critics have noted, Bacon's view of imagination was complex and ambivalent in itself.[50] On the one hand, he says that imagination, in close conjunction with reason, can produce lively "pictures of virtue and goodness," using the "ornament of words." On the other hand, the fantasy tends to pluck images from outside the realm of experience, contrary to the inductive method, which aims to discover knowledge in systematic rather than accidental ways.[51] At the same time, Bacon felt that "schools, academies, colleges, and similar bodies" were wrongly averse to innovation: "The lectures and exercises there are so ordered, that to think or speculate on anything out of the common way can hardly occur to any man."[52] He himself would freely draw on fancy in his *New Atlantis*, a work of early science fiction that rather tries to think "out of the common way" and conceives in the wondrous House of Salomon a sophisticated research laboratory far ahead of its time. Still, the scientific value of fancy remains unclear in Bacon and in early modern thinking as a whole. In a similarly noncommittal way, for example, Cornwallis concedes that fancy may lead to wisdom, albeit in a circuitous fashion:

---

47. Bacon, *Advancement of Learning*, 4:282–84.

48. Bacon, 4:285, 290.

49. Bacon, 4:282, 289.

50. For more on the mutable characterization of imagination in Bacon's corpus, see, for example, Karl R. Wallace, *Francis Bacon on the Nature of Man: The Faculties of Man's Soul: Understanding, Reason, Imagination, Memory, Will, and Appetite* (Urbana: University of Illinois Press, 1967); Eugene P. McCreary, "Bacon's Theory of Imagination Reconsidered," *Huntington Library Quarterly* 36, no. 4 (1973): 317–26; and Ronald Levao, "Bacon and the Mobility of Science," *Representations* 40 (1992): 1–32.

51. Francis Bacon, *Of the Dignity and Advancement of Learning*, in *Works of Francis Bacon*, 4:456. See also Michel Malherbe, "Bacon's Method of Science," in Peltonen, *Cambridge Companion to Bacon*, 76.

52. Francis Bacon, *The New Organon*, in *Works of Francis Bacon*, 4:89.

"Experience comes from Knowledge, Knowledge from Chaunge, Chaunge from Fantasticknesse."[53] It seems that the power of imagination was somehow connected to discovery and invention, though the connection was hard to articulate.

By bestowing the "fantastical" flaw on both Holofernes and Armado, Shakespeare elides rather than elucidates the difference between scholasticism and innovation, between the pedantic and the newfangled. This explains why imaginative idling is always a pejorative accusation no matter who is speaking in Love's Labor's Lost, and is always an infraction readily alleged against the other. The King mocks Armado; Armado faults Holofernes; Holofernes chides Armado—one man's intellectual giant is another man's "phantasime." The pervasiveness of imaginative idling suggests that imagination underlies both modern and premodern forms of knowledge making, and it suggests the probable futility of any attempt to contain or sequester fancy. At the same time, the play's linguistic pyrotechnics illustrate how a method of inquiry founded on imagination might giddily proceed. To deploy the fantasy means to invite the mind's ceaseless phantasmatic transfigurations, its frenzied rhythm of self-interruption. In toying with fancy, we can stumble upon that which is beyond our ken, that which is not yet even known to be unknown; this is why "that sport best pleases that doth least know how" (5.2.514). Yet there remains a risk that the sportive churn may be fruitless and result in lost labor. Imagination promises both unpredicted success and failure, underwrites both the desire for the new and the distrust of newness.

Shakespeare underlines the universality of fancy's lure by making all of Navarre's inhabitants into children. There happens to be an actual child in the play: Mote, a "most acute juvenal" who easily outwits the adults, tutoring his master, Armado, and trouncing Holofernes in a joust of wits ("Thou disputes like an infant," sniffs the schoolmaster) (3.1.58, 5.1.59). There is also the suggestively named Boyet (boy yet), and the pageant of Nine Worthies, in which we see Alexander the Great bullied and Hercules reduced to "a babe, a child, a shrimp" (5.2.584). In their infatuation with the ladies, meanwhile, the four courtier-scholars are said to descend into "an old infant play" (4.3.73).[54] Biron says that to see his friends in the grip of fancy is

53. Cornwallis, Essayes, N4.

54. Richard Corum addresses the play's adolescent aspect in a different context, arguing that the men deliberately sabotage their wooing attempts, not actually wanting to win what they pursue. "'The Catastrophe Is a Nuptial': Love's Labor's Lost, Tactics, Everyday Life," in Renaissance Culture and the Everyday, ed. Patricia Fumerton and Simon Hunt (Philadelphia: University of Pennsylvania Press, 1999), 271–98.

to see great Hercules whipping a gig,
And profound Solomon tuning a jig,
And Nestor play at push-pin with the boys,
And critic Timon laugh at idle toys. (4.3.162–65)

In indulging fancy, it appears, we all become children. Imagination cannot be excluded from fields of knowledge making, cannot easily be disentangled from intellectual work. In the likes of Holofernes, it aids the application and manipulation of established knowledge; in Armado, it drives the hunt for untested conceits; for the lovers, it is needed to play the games of courtship. All strive for expertise and experience using fancy's tactics: lateral logic, purposelessness industry, modish newfangledness, and the compulsive kneading of words.

FAREWELL TO FANCY

While the minor plot of *Love's Labor's Lost* describes imagination's chaotic method, the main plot considers its relation to empiricism. As we saw earlier, the play begins with the King and his men wishing to vanquish their impulses and affections. The Princess and her ladies, it turns out, are just as intent on conquest—waging a "civil war of wits" with the "bookmen" of Navarre (2.1.225–26). The very first thing the Princess says is that she dislikes canned praise. This comes in response to a reflexive compliment issued by Boyet, who remarks that Nature "did starve the general world" in "prodigally" bestowing all "graces" on his mistress (2.1.10–12). She replies,

My beauty, though but mean,
Needs not the painted flourish of your praise.
Beauty is bought by judgment of the eye,
Not uttered by base sale of chapmen's tongues. (2.1.13–16)

A flatterer who spends his wit in the praise of another merely wishes "to be counted wise" himself; he gives in excess of what is warranted, gives in order to receive (2.1.18). This is unnecessary, reasons the Princess, because beauty is adequately "bought by judgment of the eye" and does not need the "painted flourish" of words. Discreetly accepting the lesson, Boyet does better later on. When he persuades the Princess that the King is smitten with her, he is careful to lead with evidence he has garnered through "my observation, which very seldom lies":

Methought all his senses were locked in his eye,
As jewels in crystal for some prince to buy,
Who, tend'ring their own worth from whence they were glassed,
Did point you to buy them along as you passed.
His face's own margin did quote such amazes,
That all eyes saw his eyes enchanted with gazes. (2.1.241–46)

Rather than rehearse the Princess's beauty, Boyet focuses on the King's face. He does not attempt to "quote" in words what was signaled there. Rather, he merely recounts what he saw—"jewels," "crystal," "glass." "I only have made a mouth of his eye," he says innocently (2.1.251). This goes over well; the women commend him for speaking "skillfully."

Deriding the men for their barren wit, the ladies want them instead to accept that the apperception of beauty is sensual rather than linguistic and leave off unnecessary speechmaking. Words and matter must correlate, the women insist as they appraise their male counterparts. Thinking of Longueville, Maria notes that a sharp "wit" is often accompanied by a "too blunt" "will" (2.1.49). About Dumaine, Catherine observes that wit can compensate for a deficit of looks, "make an ill shape good"; if the looks are adequate, however, the wit is superfluous and tiresome (2.1.59–60). Rosaline styles Biron's "wit" at greatest length:

His eye begets occasion for his wit,
For every object that the one doth catch
The other turns to a mirth-moving jest,
Which his fair tongue—conceit's expositor—
Delivers in such apt and gracious words
That agèd ears play truant at his tales
And younger hearings are quite ravishèd,
So sweet and voluble is his discourse. (2.1.69–76)

Biron's "wit" is really a witless stream of bons mots provoked but not truly guided by what he sees. The eye feeds "objects" to the mind's "conceit," which rapidly breeds "mirth-moving jest" in response. Rather too rapidly, in fact—the engine of Biron's imagination here appears as robotically mechanical as Armado's spontaneous word mintings and Holofernes's reflexive Latinisms. The effect is amusing in a juvenile sort of way, appealing to "younger hearings" and those "agèd ears" inclined to "play truant." But these automatic responses of an idle imagination can be properly pleasing only when they are accompanied by sensory perceptiveness. The sort of fancy that will be required from now on will be newly grounded in empirical experience, not derived from ancient conjectures and scholastic proofs.

Epistemologically, the women privilege experience based on direct observation, predicting the method of induction later described by Bacon's *New Organon*, in which scientific knowledge does not originate from "the most general axioms" but rather "from the senses and particulars, rising by a gradual and unbroken ascent, so that it arrives at the most general axioms last of all."[55] In Aristotelian epistemology, things are known by way of "experience"—that is to say, by way of commonsense notions consolidated through consensus into "a universal statement of how things are, or how they behave." Under this scheme, collectively accepted perceptions of others count above those of the individual, even if they have not been personally experienced.[56] What we see in *Love's Labor's Lost* is, again, nicely reflective of Shakespeare's transitional historical moment—namely, dissatisfaction with experience that is isolated from empirical evidence, experience that denies the observational realities of the here and now and the experiential sensations of the observer. Rather, there is an intuition, strongest among the canniest persons of the play, that empirically sourced experience can substitute for knowledge gathered through books and conventional wisdom. "How hast thou purchased this experience?" asks Armado of swift-witted Mote after the boy makes an impressively knowing speech about the rituals of courtship. "By my penny of observation," replies the page (3.1.22–23).

Those who have not yet learned to remake experience in this novel way will be required to do so; this is the real education received by the male scholars. In order to please their ladies, the men will rehabilitate their ways of thinking and imagining. As Biron announces in act 4, they decide to renounce reading in favor of looking, to exchange books for eyes:

Oh, we have made a vow to study, lords,
And in that vow we have forsworn our books.
For when would you, my liege, or you, or you,
In leaden contemplation have found out
Such fiery numbers as the prompting eyes
Of Beauty's tutors have enriched you with?
Other slow arts entirely keep the brain,
And therefore, finding barren practicers,

---

55. Bacon, *New Organon*, 4:50.
56. Peter Dear, *Discipline and Experience: The Mathematical Way in the Scientific Revolution* (Chicago: University of Chicago Press, 1995), 22.

Scarce show a harvest of their heavy toil.
But love, first learned in a lady's eyes,
Lives not alone immurèd in the brain,
But with the motion of all elements
Courses as swift as thought in every power,
And gives to every power a double power,
Above their functions and their offices. (4.3.313–27)

From women's eyes this doctrine I derive:
They sparkle still the right Promethean fire.
They are the books, the arts, the academes,
That show, contain, and nourish all the world,
Else none at all in aught proves excellent. (4.3.345–49)

This consecration of "women's eyes" sounds like the institution of a new academic charter, almost. The speech endorses the mandate of early modern empiricism, "to dissect nature, not merely to dissect language," and it reimagines the traditional curricular structure.[57] Comparison with an earlier draft of this passage reveals that Shakespeare's thinking on these subjects was evolving. The shorter quarto version says that "universal plodding poisons up / The nimble spirits in the arteries" (4.3.299–300); there is no mention of the senses. In the later folio, learning extends beyond the embodied mind: it is not "immurèd in the brain," inhering rather in the "motion of all elements," running "as swift as thought in every power." Biron goes on to detail how each of the senses is heightened—eyes, ears, feeling, tongue. As the diction of mobility implies—"slow," "motion," "swift," "courses"—the emphasis is on movement rather than the stillness of contemplative "living art." Another revision that Shakespeare made was to add multiple references to the disciplinary "arts." The quarto speech says,

Learning is but an adjunct to ourself,
And where we are, our learning likewise is.
Then when ourselves we see in ladies' eyes
With ourselves—
Do we not likewise see our learning there? (4.3.308–12)

These lines present knowledge as solipsism, circumscribing it within the perceptual apparatus of the individual. The later draft, however, situates knowl-

---

57. Wallace, *Francis Bacon*, 6.

edge amid social structures, underlining the institutionalized shape of academic learning. Compare the earlier drafted line—"the ground, the books, the academes"—where "ground" emphasizes existing intellectual foundations, with the folio equivalent, which emphasizes that learning is a *techne* or "art"— "the books, the arts, the academes" (Q 4.3.297, F 4.3.347). Just as these courtier-scholars could only swear their vows of study after committing to break them, it seems they can only praise books, arts, and academes by reframing learning as an activity rather than an outcome, as an intellectual journey rather than "leaden contemplation." In its updated form, Biron's laudation of "women's eyes" makes a place for fancy among the brain's various "slow arts."

This realization appears to help the male lovers graduate beyond the distortions of "painted rhetoric" (4.3.233). After the embarrassments of the Muscovite masque, Biron in particular is ready to renounce quaint costumes and putrid learning:

> Taffeta phrases, silken terms precise,
> Three-piled hyperboles, spruce affectation,
> Figures pedantical—these summer flies
> Have blown me full of maggot ostentation. (5.2.407–10)

Both extremes of barren wit are rejected in the same breath: the precision of pedantry and the "ostentation" of "affection." The first surprise that comes at the end of *Love's Labor's Lost*, a play rather filled with "taffeta phrases" and "silken terms," is that its protagonists disavow the rhetorical fripperies with which they have thus far entertained themselves and us. When news arrives of the French King's death, none other than "honest plain words" seem permissible. Biron apologizes for his waggish antics, calling love

> all wanton as a child, skipping and vain,
> Formed by the eye, and therefore, like the eye,
> Full of straying shapes, of habits, and of forms,
> Varying in subjects as the eye doth roll
> To every varied object in his glance. (5.2.747–51)

The latter three lines predict Theseus's speech on imagination—there it is the poet, rather than the lover, whose eye, "in a fine frenzy rolling," fashions the "forms of things" into intelligible "shapes" (5.1.12–16). But "strange shapes" of wanton fancy receive no quarter from the women. Shrugging off their previous flirtation as a "pleasant jest," a mere "lining of the time," the now Queen of France suggests that Ferdinand betake himself to "some forlorn and naked hermitage" for more study (5.2.781). In the second twist, it turns out that imagination is going to be outlawed too—at least in its present form. The path toward this new, reconfigured fancy lies via the sensing body, and so the Queen and her attendants shoo the men toward an "austere insociable life" (5.2.785). Sublimated by "frosts and fasts," the "fruitful brain" will have to attend to "groaning wretches" and try to soothe "sickly ears" (5.2.787, 833, 838, 849).

Thus, the comedy ends fruitlessly, as might any idle fantasy. The ending may be read in league with what Cynthia Lewis calls the play's insistence on the "fact of mortality."[58] It insinuates the double reality of the play: as Louis Montrose writes, death reveals Navarre's "imaginative world of study, dreams of fame, games, and courtship" to be a ludic space that has been deliberately and artfully constructed to exclude politics, war, and death.[59] This juxtaposition of mortality and games, striking though it may seem, was not so strange for Montaigne. "There is nothing wherewith I have ever more entertained my selfe," he wrote, "than with the imaginations of death." This is not less but rather more so the case when he is "amongst faire Ladies, and in earnest play." In the midst of such disports, Montaigne cannot help but dwell on "the remembrance of some one or other, that but few daies before was taken with a burning feuer, and of his sodaine end, comming from such a feast or meeting where I was my selfe, and with his head full of idle conceits, of love, and merry glee."[60] The passage is taken from his essay "That to Philosophie, is to learne how to die"; it seems almost to imply that idle toying and presentiments of death are each alike paths to wisdom. Following Montaigne, Shakespeare's play's ending is neither a wholesale espousal of "merry glee" nor a grim statement of morbid inevitability. Rather, it seems to call for a new state of affairs, one in which bodily corporeality and the metaphysical mind are bridged. That it is the task of the Navarrese men to erect this bridge by transforming themselves implies that the imagination is due for a sort of transformation too—it must become alert to things outside itself, to nature rather than preconceived notions about nature.

At the end, the play does not so much undo its work as defer the completion of that work: its women and men do not part ways forever, only for a time. A temporary farewell to fancy and a literary experiment in its own right, *Love's Labor's Lost* incorporates both the playful exuberance and the subtle

58. Cynthia Lewis, "'We Know What We Know': Reckoning in *Love's Labor's Lost*," *Studies in Philology* 105, no. 2 (2008): 261.

59. See Louis Montrose, "'Sport by Sport O'erthrown': *Love's Labour's Lost* and the Politics of Play," *Texas Studies in Language and Literature* 18, no. 4 (1976): 543.

60. Michel de Montaigne, "That to Philosophie, is to learne how to die," in *Essayes*, 34.

seriousness with which the early moderns imagined the use and disuse of imagination. Shakespeare's comedy proceeds from the existing premises of early modern epistemology—in which monastic seclusion vies with courtly sociality, pedantry with gentility, syllogism with experience, imitation with invention—toward the derivation of its own conclusions about idle imagination. Chief among these is that fancy can be a byword for innovation, its promises and pitfalls. As Shakespeare's children of fancy go about their wooing, rhyming, duping, and jesting, they intimate the fantasy's power to foil mental habits, academic structures, and scholarly norms. Entertaining and then frustrating its audience, the comedy shows that the Renaissance instinct to systematize knowledge making in turn broached questions about the intellectual stakes of imaginative play, as well as the attractiveness of toys.

CHAPTER 3

Of Atoms, Air, and Insects
Mercutio's Vain Fantasy

If there was in sixteenth-century English a descriptor of imagination that could rival the word *idle* in both ubiquity of usage and complexity of connotation, that word might be *vain*. We saw in the previous chapter how the notion of idleness encodes ambiguities surrounding the epistemological potential of fantasy and its relation to different knowledge traditions and methods. The idea of vanity, in contrast, engages the physical content of fancy, the materiality and substantive nature of phantasms—or, to be more accurate, their immateriality. In *Romeo and Juliet*, Mercutio suggests that "vain fantasy" is empty of value:

> I talk of dreams,
> Which are the children of an idle brain,
> Begot of nothing but vain fantasy,
> Which is as thin of substance as the air
> And more inconstant than the wind.[1]

Mercutio's opinion is that fancies are hollow, devoid of matter as well as significance. It is an opinion entirely consistent with prevailing cultural prejudices.

---

1. William Shakespeare, *Romeo and Juliet*, 1.4.94–98. Shakespeare references are from *The Norton Shakespeare*, ed. Stephen Greenblatt et al. (New York: W. W. Norton, 2016), hereafter cited parenthetically.

Nevertheless, the formulation carries some internal tensions. As Mercutio would have us believe, mental images are "thin of substance"—but not so thin that they cannot breed. They are close to "nothing" yet are able to move like the changeable wind. Though dreams exist in the "brain," they are also misbegotten offspring, having a physical origin, capable of being born. The peculiar physics of Mercutio's "vain fantasy" arguably raises more questions about imagination's material nature than it answers. What do phantasms consist of, and how do they come into being?

In religious and spiritual discourse, the supposed vanity of imagination is connected to the vacuity of earthly life. There is evidence to suggest, though, that this ideologically unproblematic view had nuances, nuances that were mobilized with precision within sixteenth-century theological debates and were separately enhanced by the surging interest in atomist philosophy. Classical materialism, we know, was rediscovered in the fifteenth century, after which it slowly came to pervade Renaissance culture and science. The corpuscular universe posited in particular by Lucretius would be taken up by early modern natural philosophers such as Daniel Sennert and Pierre Gassendi in the service of solving problems in kinetic theory, biology, and optics. Along with the philosophy of Lucretius's *De rerum natura*, though, came its striking figurative tropes. Though the idea of a Christian cosmos composed of atoms presented ideological difficulties, the literary and imaginative possibilities released by classical materialism were rich, as we see in Elizabethan representations of mental imagery. Fantasy becomes a swarm of whirling motes, a breeding ground for worms and flies, the intrusion of fine substances into porous bodies. This characterization, which coexisted with traditional connotations of vanity, renders imagination both carnal and spiritual, gross and sheer, vital and morbid.

Mercutio's references to thin matter, begotten broods, and inconstant wind gesture to this increasingly paradoxical ontology of fantasy, as does the wider imagery of *Romeo and Juliet*. With its fumes, webs, and mists, the play endows its eponymous lovers with an imaginative receptivity that counters Mercutio's more critical view. Sensitized constantly to their imminent fleshly disintegration through dreams and prophetic fantasies, Juliet and Romeo interrogate the perceptual distinction between substance and show; they train their attention on thin substances and so explore the disjunction between the material reality of the cosmos and the perceptual limitations of the mind. The most explicit culmination of the tragedy's materialist explorations comes in Mercutio's quarrel with Romeo over the nature of "nothing," during which Mercutio delivers his virtuosic rhetorical tribute to Queen Mab, the imaginary fairy who ushers dreams into the minds of slumbering lawyers and ladies. Though the speech is meant to illustrate decisively the vanity of fantasy, Mab evades Mercutio's grasp soon after he calls her into being. She exposes the function of figurative language in natural philosophy, suggesting that the subtle tangibility of phantasms is a fiction necessary for scientific explanation.

ATOMS, AIR, AND INSECTS

Thomas Cooper's 1578 thesaurus defines *phantasma* as "a vaine vision," testifying to the intimate association between vanity and fantasy in sixteenth-century thinking.[2] Following the idea in Genesis that "the wickednes of man was great in the earth, and all the imaginations of the thoughts of his heart were onely euill continually,"[3] the image-making mind invokes the prideful vanity of humankind, as well as the *vanitas* or fundamental emptiness of earthly existence. In accordance with this, sermons from the period berate believers for "talking with the vanities of the world, or with the foolish imaginations of our owne hearts"; they remind parishioners that the wicked are happy only in "vayne imagination, and they do but dreame."[4] This being said, religious discussions of the period belie a deep interest in the mysterious nature of imagination's substantiality. In post-Reformation debates about transubstantiation especially—which were in effect vigorous metaphysical arguments over the nature of matter—the vanity of fantasy is mobilized in various ways.[5] In these fierce semiotic disputes over the nature of Christ's presence at the holy supper, the idea of imagination stands variously for the idolatrous worship of material objects, the absence of substance, the flawed operations of the embodied mind, and a signifier lacking a signified. A brief look at this discourse indicates the many ways in which the materiality of phantasms was thought about in the late sixteenth century.

In the simplest sense, imagination is a synonym for error. Thus, one English translation of the Reformer Peter Martyr Vermigli terms the notion of

---

2. Thomas Cooper, *Thesaurus linguae Romanae & Britannicae* (London, 1578), Ccccc5r, s.v. "phantasma."

3. *The Geneva Bible: A Facsimile of the 1599 Edition* (Ozark, MO: L. L. Brown, 1990), Genesis 6:5.

4. Henry Bull, *Christian praiers and holie meditations* (London, 1578), 22; John Calvin, *Sermons of Master Iohn Calvin, vpon the booke of Iob*, trans. Arthur Golding (London, 1574), 415.

5. See Sophie Read, *Eucharist and the Poetic Imagination in Early Modern England* (Cambridge: Cambridge University Press, 2013), 6–8, for a discussion of how the Eucharist provoked "new ways of understanding signification itself" (3). See also Kimberly Johnson on the mobilization of "the material" in seventeenth-century eucharistic poetics; *Made Flesh: Sacrament and Poetics in Post-Reformation England* (Philadelphia: University of Pennsylvania Press, 2014), 29.

transubstantiation a baseless "toye and fantasie."[6] To worship such an idea, the Protestant author William Tyndale said, is a sort of "idolatry"[7]—all the more so if the idea itself is about idolatry, about the wrongful idolization of the Mass. The consecrated meal of the Eucharist is "a phantasticall thing" too, for, as Catholics maintained, the blessed bread is not really bread; it only seems so "to the outward sense."[8] This "empty imaginatiue forme of bred" cannot function as a sign of holy communion—if the food is not actually food, is not even itself, asks John Calvin, how can it symbolize some other thing?[9] To give credence to this fantastical bread is furthermore akin to giving Jesus "a phantasticall body"—an old heresy from the church's early days.[10] Wanting to do away with all these fantasies, Protestants sought to remake the Eucharist, replacing the carnal presence of Christ with a spiritual remembrance of his sacrifice on the cross.[11] But this, claimed the other side, reduces the sacrament to a mere *"imagination, . . . conceit and fantasie."*[12] It bears noting that in this complex debate, "imagination" is never a satisfactory ontological category. The Lord can be invisible, spiritual, remembered, or symbolized; what he cannot be is fantastical.

While phantasms often stand for immateriality, they also serve as reminders that the human mind is embodied, that we are predisposed to think in sensory terms. Catholic and Protestant commentators alike stress that it is wrong to engage "grosse imaginacions and carnal fantasies" about eating Christ's body.[13] Nicholas Sander devotes a chapter in his *Supper of our Lord* to listing the repugnant fantasies that must be held at bay when taking communion: the notion that "we doe eate" a "body of a man"; that the body has been broken into pieces, "one taking the shoulder, an other the legg"; that the body and the bread coexist, as though it were possible to "eate that immortall and glorious flesh of Christ, with bakers bread."[14] Carnality is invoked in another sense, less squeamishly but maybe more shockingly, by the Dutch theologian

---

6. Peter Martyr Vermigli, *A discourse or traictise of Petur Martyr Vermilla Florentine*, trans. Nicholas Udall (London, 1550), 25v.

7. William Tyndale, *The whole workes of W. Tyndall* (London, 1573), 161.

8. Thomas Beard, *A retractiue from the Romish religion* (London, 1616), 500.

9. John Calvin, *The institution of Christian religion*, trans. Thomas Norton (London, 1561), 123v.

10. Thomas Cranmer, *An aunswere by the Reuerend Father in God Thomas Archbyshop of Canterbury* (London, 1580), 183. See also William Perkins, *A reformed Catholike* (Cambridge, England, 1598), 342: "They have plainly turned the true God into a phantasie of their own."

11. See William Rainolds, *A treatise conteyning the true catholike and apostolike faith of the holy sacrifice and sacrament ordeyned by Christ at his last Supper* (Antwerp, 1593), 129: "Albeit the thing signified be corporally absent, yet a faithful imagination and sure faith reneweth or remembreth that worke once done."

12. Rainolds, 312.

13. Thomas Cranmer, *A defence of the true and catholike doctrine of the sacrament of the body and bloud of our sauiour Christ* (London, 1550), 80v.

14. Nicholas Sander, *The supper of our Lord* (Leuven, Belgium, 1565), 85v, 86v.

Philips of Marnix. The idea that Christ is in the bread, says the author mockingly, would suppose that he is a "litle nothing, hanging by a small silke thred" within the dough. Are we to believe that he is a "litle incomprehensible winde, or moth, which flieth hence awaye in the aire"? That the godhead is like those "manie litle wauering things, alwayes shaking and flying in the aire," which the Greeks called "*Atomi*"?[15] Here, the heresy of transubstantiation is conflated with the faithless "fantasie" of Epicureanism; the Son of God becomes a flitting "moth"—or "mote," as it might have been pronounced then. Sardonically pretending to apply physics to divinity, the author stresses that the two should not be stupidly confused.

As Thomas Cranmer explains it, the stakes of understanding the Holy Communion correctly are literally astronomical. The archbishop of Canterbury scolds his opponents for "phantasiyng substance" as a thing unto itself that is separable from material "accidents." To do this is tantamount to confounding all things together, to mixing "heauen and earth," to saying that "all flesh is one flesh, and all substances one substance," and so to making an indiscriminate "hotche potche" of "the sunne and the moone, of a man and a beast, of fish and flesh, betwene the body of one beast and an other, one herbe and an other, one tree & an other, betwene a man and a woman."[16] Chaos results when "substance" is thus unmoored. Implicit in Cranmer's account is the power of imagination to make or unmake a coherent cosmos. If the material architecture of the universe is "phantas[ied]" erroneously, wrongly construed, then everything slides into meaninglessness. Thus, it would appear that a fantasy is both an idea of no substance—no value, no truth—and an idea too much grounded in substance—too carnal, too gross.

Like Marnix's silken thread of atoms, Cranmer's cosmic hodgepodge evokes the atomist philosophies of antiquity. As Stephen Greenblatt, Gerard Passannante, and others have shown, the materialist philosophies of Democritus and Epicurus garnered renewed attention in European culture after Poggio Bracciolini's discovery in 1417 of Lucretius's six-part philosophical verse treatise *De rerum natura*.[17] In vivid language, the poem explores the implications of its central premise—namely, that the world is made up of atoms, minute particles swirling in a void. It contends that the world was not created by God, that

15. Philips van Marnix van St. Aldegonde, *The bee hiue of the Romishe Church* (London, 1579), 107r.
16. Cranmer, *Defence of the true*, 275.
17. Stephen Greenblatt, *The Swerve: How the World Became Modern* (New York: W. W. Norton, 2011); Gerard Passannante, *The Lucretian Renaissance: Philology and the Afterlife of Tradition* (Chicago: University of Chicago Press, 2011). See as well chap. 6 in Howard Jones, *The Epicurean Tradition* (London: Routledge, 1989); and Alison Brown, *The Return of Lucretius to Renaissance Florence* (Cambridge, MA: Harvard University Press, 2010).

there is no afterlife because the soul is material and therefore mortal, and that
pleasure is the highest good. Needless to say, these tenets hold profound ramifications for any system of understanding the foundations of life, the composition of the cosmos, countless natural phenomena, and of course the workings
of the human mind. An explicit affirmation of atheism, hedonism, and the tran-
sience of the soul, Epicureanism was incompatible with early modern Christian
metaphysics, adherents of which dismissed its "ridiculous blasphemy."[18]
Among natural philosophers, its ideas provoked necessarily strategic responses and cautious adaptations. Lucretius was not the only source of atomist ideas circulating in Renaissance Europe, it should be noted. The early
moderns would have encountered the idea of small particles via Aristotle's
minima—these were not atoms, exactly, but rather "small portions of matter
that enabled the miscibles to mingle and interact"—which were in turn inspired by Democritus. Thomas Harriot, who believed firmly that nothing
comes from nothing—*ex nihilo nihil fit*—also dabbled in atomism, as did Girolamo Fracastoro, who employed the notion of seeds to explain phenomena
such as magnetism and evaporation.[19] Nonetheless, the imaginative sophistication of *De rerum natura* sets it apart. As Passannante has argued, Lucretius's
analogies had a foundational impact on Renaissance intellectual history,
furnishing postclassical thinkers with a means of conceiving new forms of
textual transmission and literary tradition.[20] In a comparable way, we will
see, Lucretian tropes permeated the contemporary conception of imagination. Refracted through atomist tropes, the image-making faculty was itself
imagined in terms of motes, worms, and winds.

Key among Lucretius's images is the one meant to help us visualize atoms
themselves: "Do but apply your scrutiny whenever the sun's rays are let in and
pour their light through a dark room: you will see many minute particles min-

18. Abraham Fraunce, *The lawiers logike* (London, 1588), 17v.

19. Norma E. Emerton, *The Scientific Reinterpretation of Form* (Ithaca, NY: Cornell University Press, 1984), 87; Hilary Gatti, "The Natural Philosophy of Thomas Harriot," in *Thomas Harriot: An Elizabethan Man of Science*, ed. Robert Fox (Aldershot, England: Ashgate, 2000), 70–71. See as well Christoph Meinel, "Early Seventeenth-Century Atomism: Theory, Epistemology, and the Insuffi-ciency of Experiment," in *The Scientific Enterprise in Early Modern Europe: Readings from "Isis,"* ed. Peter Dear (Chicago: University of Chicago Press, 1997), 176–79; and Mary Thomas Crane, *Losing Touch with Nature: Literature and the New Science in Sixteenth-Century England* (Baltimore: Johns Hopkins University Press, 2014), 81.

20. See Passannante, *Lucretian Renaissance*, esp. chap. 3: "The poetry of materialism transformed not only the principles of modern science, but also an understanding of intellectual debt, the tech-nologies of transmission and the idea of the unity and conservation of knowledge" (122). Meinel agrees that whereas the empirical basis of Lucretian atomism remained weak, its "poetic and imagi-native qualities" were not; "Early Seventeenth-Century Atomism," 176, 185.

gling in many ways throughout the void in the light itself of the rays."[21] This image of particles illuminated in a sunbeam recurs in early modern texts: Philemon Holland's translation of Plutarch's *Morals*, for example, defines "atomi" as "motes in the Sunne beames."[22] With many more concrete and memorable images, Lucretius explains how various cosmological phenomena can be explained by the behavior of atomic particles. Transported in mists and heat, the corpuscles draw life seemingly out of death, as when worms appear in dung and flies around carcasses. Atoms cannot be created or destroyed, merely transmuted from one configuration into another: sticks turn into fire; animals are absorbed into the other animals that eat them; flesh rots (2.873–85). Early modern natural philosophers would investigate these ideas further. Daniel Sennert, for instance, studied biogenesis at length, concluding with the aid of atomist principles that what appears to be spontaneous generation is brought about by "smallest Bodies or Petty Atomes," which, though invisible or dormant, contain "a vital principle."[23] These living seeds, dispersed through rain and dew and carried in plants and animals, germinate in the right conditions, producing worms, caterpillars, and other creepers.[24] With its many examples of life spawned through putrefaction and procreation in both the earth and the air, Sennert's text has all of the color of Lucretius's poem, writhing with beetles, fleas, and eels.

Another striking set of images derives from the Lucretian theory of perception, which is built on the notion that things in the world constantly shed fine particulate films, or *eidola*: "Amongst visible things many throw off bodies, sometimes loosely diffused abroad, as wood throws off smoke and fire heat, sometimes more close-knit and condensed, as often when cicadas drop their neat coats in summer, and when calves at birth throw off the caul from their outermost surface, and also when the slippery serpent casts off his vesture amongst the thorns" (4.54–61). These discarded simulacra are sufficiently fine in texture that they can travel at great speeds, entering into the perceiver unnoticed. They "penetrate through the interstices of the body, and awake the thin substance of the mind [*tenvem animi naturam*] within," for "the mind is itself thin and wonderfully easy to move" (4.729–30, 747–8). When this happens, we seem to see, imagine, or dream of the things themselves. Mental

---

21. Lucretius, *On the Nature of Things*, trans. W. H. D. Rouse, rev. Martin F. Smith, Loeb Classical Library 181 (Cambridge, MA: Harvard University Press, 2002), 2.114–18. Subsequent references appear in the text.

22. Plutarch, *The philosophie, commonlie called, the morals*, trans. Philemon Holland (London, 1603), Zzzzz1v.

23. Daniel Sennert, *Thirteen books of natural philosophy* (London, 1660), 207.

24. Sennert, 185, 211, 204.

vision and actual vision are therefore effectively identical processes; it is only that mental images are considerably finer in texture and so elude detection by the eye. As before, metaphors render these wafting skeins wonderfully palpable. Lucretius compares them to animal skins and afterbirths: the "neat coats" left behind by cicadas, the calf's discarded "caul," and the serpent's molted "vesture." As a theory of cognition, Epicurean simulacra gained little traction in the early modern period: Sennert writes that the notion of atomic "*Skins and Barks*" was long ago rejected.[25] Still, it speaks to the power of this imagery that it is taken up by atomism's detractors. The supernaturalist Pierre le Loyer does so in his *Treatise of specters or straunge sights*, for example, denying that atoms can enter the mind undetected and somehow reorganize themselves into intelligible forms after the fact. He first quotes Lucretius's tropes— the caterpillars and snakes that "do leaue their spoiles in the hedges or bushes," the "thinne and slender skin" carried in by "little creatures" from "their dammes belly" at birth—and then refutes the argument using those same metaphors.[26] "After-burthens" and other bodily "spoils," he notes, could hardly be confused for "an Image" or "impression" of the creature from which they originate. Such remnants are "superfluous," "no other then a thinne slender skinne." They are not identical to the animals' bodies, for then "of one body there should be two made; which were a straunge thing, and altogether abhorring from nature."[27]

It is in a figurative rather than epistemological sense that Lucretian materialism makes its mark on conceptions of imagination. Aristotelian faculty psychology was not displaced by Epicurean ideas; yet Renaissance representations of the fantasy exhibit traces of the latter. Perhaps the most startling such depiction is the chamber of imagination depicted in *The Faerie Queene*. Housed in the Castle of Alma is "Phantastes," a figure with "a sharpe foresight, and working wit, / That neuer idle was, ne once would rest a whit."[28]

His chamber was dispainted all with in,
    With sondry colours, in the which were writ
    Infinite shapes of thinges dispersed thin;
    Some such as in the world were neuer yit,
    Ne can deuized be of mortall wit;

25. Sennert, 385.
26. Pierre le Loyer, *A treatise of specters or straunge sights*, trans. Zachary Jones (London, 1605), 27r. Le Loyer's work on specters went through multiple French and English editions starting in 1586.
27. Le Loyer, 30v.
28. Edmund Spenser, *The Faerie Qveene*, ed. A. C. Hamilton, text ed. Hiroshi Yamashita and Toshiyuki Suzuki (New York: Longman, 2001), 2.9.49.

> Some daily seene, and knowen by their names,
> Such as in idle fantasies doe flit:
> Infernall, Hags, *Centaurs*, feendes, *Hippodames*,
> Apes, Lyons, Aegles, Owles, fooles, louers, children, Dames.
>
> And all the chamber filled was with flyes,
>   Which buzzed all about, and made such sound,
>   That they encombred all mens eares and eyes,
>   Like many swarmes of Bees assembled round,
>   After their hiues with honny do abound:
>   All those were idle thoughtes and fantasies,
>   Deuices, dreames, opinions vnsound,
>   Shewes, visions, sooth-sayes, and prophesies;
> And all that fained is, as leasings, tales, and lies.[29]

The first stanza culminates in a catalog of fantastical beings, some of which are mythic ("*Centaurs*," "feendes"), some bestial ("Lyons," "Owles"), and some human ("fools," "children"). The infinite "shapes" are practically weightless: they "flit" around the room, "dispersed thin." In the second stanza, what were previously isolated flutterings gather into a droning swarm, giving rise to a second list—"deuices," "dreames," "opinions," "shewes." The words crowd the page, seemingly multiplying before our eyes, while the accumulating sibilants raise a mounting hum: "flies," "eyes," "hives," "lies." These apian insects—they swarm like bees but are not bees, exactly—pictorialize a surface completely overrun, a place filled to bursting. Spenser spatializes cognition, makes thoughts into obfuscating mass. As we saw with Shakespeare's anatomizing sonnets in chapter 1, the ophthalmological plausibility of Phantastes's condition makes Spenser's representation all the more arresting. Accordingly to medical books of the time, cataracts and myopia really could cause people to see small bodies— such as "flies, haires, or threeds of a spider web"; "little gnattes: and sometimes little atoms"; "motes which flie in the aire"—floating before their eyes.[30]

Putrescence and pullulation, meanwhile, provide authors with a means of accounting for imagination's seemingly unnatural fertility, the way the faculty

---

29. Spenser, 2.9.50–51. Compare Spenser's sharply staring Phantastes with John Davies's depiction of "Phantasie" in *Nosce teipsum* (London, 1599), 47: "A thousand Dreames phantasticall and light, / With fluttering wings do keepe her still awake."

30. See, respectively, André du Laurens, *A discourse of the preseruation of the sight*, trans. Richard Surphlet (London, 1599), 56; Le Loyer, *Treatise of specters*, 91v; and Jacques Guillemeau, *A treatise of one hundred and thirteene diseases of the eyes*, trans. Richard Banister (London, 1622), C5r. Guillemeau's treatise was earlier published as *Traité des maladies de l'oeil* (Paris, 1585).

appears to contravene the principle that nothing can come from nothing. For example, Thomas Nash's *Terrors of the night* portrays the fantasy as a sort of miasma: just as "slime and durt in a standing puddle, engender toads and frogs, and many other unsightly creatures," so are "many misshapen obiects" born "in our imaginations."[31] The brain had vermicular associations: Galen had thought the passageway connected to the cerebellum wormlike in appearance; taking his lead, early modern physicians such as Ambroise Paré write about the cerebral "worm" and how it resembles "those thicke white wormes which are found in rotten wood."[32] From here it is not too far a stretch to see fantasies as wriggling larvae. John Lyly writes that "litle things catch light mindes, and fancy is a worme," and Ben Jonson names the crafty servant of *Every Man in His Humor* "Brainworm"—a man whose "nimble soul" works by "wak[ing] all forces of my fant'sy."[33] Shakespeare too was aware of both the miasmic and entomological associations of fantasy. In his *Rape of Lucrece*, a vermicular inching motion is offered as an image of sympathy when Lucrece's maid is moved by her mistress's sorrow. The narrator notes of women's minds,

> Their smoothness, like a goodly champaign plain,
> Lays open all the little worms that creep;
> In men, as in a rough-grown grove, remain
> Cave-keeping evils that obscurely sleep.
> Through crystal walls each little mote will peep. (1247–51)

Whereas the male interior is a "cave" or "rough-grown" grove, the female one is smooth, clear, and open, a "goodly champaign plain" that lays bare the creeping worms and peeping motes therein. The catachresis joins the crystalline to the organic, glass and dirt, producing a striking conflation of the beautiful and the grotesque. Related associations pervade the plays: in *King Lear*, we find Goneril refusing to indulge her aging father's "each buzz, each fancy, each com-

---

31. Thomas Nash, *The terrors of the night* (London, 1594), C2v. Joan Ozark Holmes has proposed that Nash might be a source for Shakespeare's speech on Mab, as both dismiss dreams as idle inventions of the phantasia. "No 'Vain Fantasy': Shakespeare's Refashioning of Nashe for Dreams and Queen Mab," in *Shakespeare's "Romeo and Juliet,"* ed. Jay L. Halio (Newark: University of Delaware Press, 1995), 51.

32. Ambroise Paré, *The workes of that famous chirurgion Ambroise Parey,* trans. Thomas Johnson (London, 1634), 168. French editions of Paré's medical texts were available in different editions during the late sixteenth century, including the collected *Oeuvres* published in 1575.

33. Ben Jonson, *Every Man in His Humor* (F), in *The Cambridge Edition of the Works of Ben Jonson,* ed. David Bevington, Martin Butler, and Ian Donaldson, vol. 4 (Cambridge: Cambridge University Press, 2012), 4.5.6–7.

plaint" (1.4.295), while in *Pericles* Gower instructs us to imagine the maritime adventures of the play's characters by saying, "Like motes and shadows see them move a while" (4.4.21).

By the mid-seventeenth century, atomism would be subsumed into the mechanical philosophy of Descartes and Gassendi. Yet images from *De rerum natura* continue to appear in literary works. Lucretius's motes suspended in sunlight, for example, appear in Milton's "Il Penseroso," where the speaker wishes to dispel a cloud of fancies:

> Dwell in som idle brain,
>   And fancies fond with gaudy shapes possess,
> As thick and numberless
>   As the gay motes that people the Sun Beams,
> Or likest hovering dreams
>   The fickle Pensioners of *Morpheus* train.[34]

Not unlike Mercutio, Milton associates dreams with motes: both are "thick," "numberless," and "hovering." More extensively than Milton, Margaret Cavendish would explore the brain-hive in her *Poems and Fancies*, where she calls mental figments "small *Gnats*" that "buz in the *Braine*":

> The *Head* of *Man* just like a *Hive* is made,
> The *Braine*, like as the *Combe's* exactly laid.
> Where every *Thought* just like a *Bee* doth *dwell*,
> Each by it selfe within a parted *Cell*.[35]

A natural philosopher herself, Cavendish explored but eventually rejected atomism, unable to accept that a body could be made up of "a swarm of Bees"— that Nature was "a Beggars coat full of lice."[36] Even so, she acknowledged that although corpuscular physics was not "serious Philosophy," it could certainly be a fertile source "for a Poetical fancy"—utilizable in imaginative poetry, if not scientifically plausible. Although Milton and Cavendish are writing at a different moment in the history of science from that of Shakespeare, they can be seen to represent the culmination of poetic trends begun in the previous

34. John Milton, *Complete Shorter Poems*, ed. Stella P. Revard (Malden, MA: Wiley-Blackwell, 2009), 53.

35. Margaret Cavendish, "Similizing the Head of Man to a Hive of Bees" and "Similizing Fancy to a Gnat," in *Poems, and fancies* (London, 1653), 149, 151.

36. Margaret Cavendish, *Observations upon experimental philosophy* (London, 1666), 142.

century. Atoms would continue to provide an occasion for poetic invention as much as a means of predictive scientific discovery.

Sennert too thought it "mad" to think "that this beautiful Theatre of the world was made by a blind and fortunate concourse of Atomes."[37] Wishing to reconcile pagan philosophy with a theocentric universe, early modern natural philosophers hypothesized that the material, the occult, and the divine could coexist.[38] Giordano Bruno's corpuscular philosophy, for example, supposed that atoms were "vitalistic," possessing an "internal form-making nucleus of light, energy, or soul."[39] Gassendi maintained that atoms are creations of God, that they are finite in number, and that their seemingly spontaneous swerves are in fact guided by divine purpose.[40] The contemporary cultural conception of imagination is of a piece with this indeterminate, controversial physics, a physics that suggests that bodies are airy and air loaded with unseen cargo. It is possible to suppose that insubstantial phantasms are weighted with matter, subject to physical laws of transmutation. Accordingly, the tropes we see in representations of imagination—motes, worms, films—alternately present the image-making faculty as a site of decomposition, a hive of activity, or a cloud of particles. These tropes show that materialism offered new ways of thinking about the still-mysterious nature of phantasms: the way they crowd the mind, arrive seemingly out of nowhere, are jostled by external stimuli. Caught between Epicurean and Christian ethics, "vain fantasy" is a thing of paradoxes—vacuous and overfull, barren yet swollen with life, fecund and decaying, all abuzz with nothing.

Substance and Show

The real anxiety, perhaps, is not so much that fantasy is emptily "vain" but rather that it might not be. *Romeo and Juliet* is fascinated by the fearsome power of spontaneous and instantaneous ideation, of potentially fatal notions delivered seemingly out of thin air and liable to multiply with abandon. This fasci-

---

37. Sennert, *Thirteen books*, 48.

38. See Adrian Streete, "Lucretius, Calvin, and Natural Law in *Measure for Measure*," in *Shakespeare and Early Modern Religion*, ed. David Loewenstein and Michael Witmore (Cambridge: Cambridge University Press, 2015), 131–54; and Keith Hutchison, "What Happened to Occult Qualities in the Scientific Revolution?," in Dear, *Scientific Enterprise*, 86–106.

39. Hilary Gatti, *Giordano Bruno and Renaissance Science* (Ithaca, NY: Cornell University Press, 1999), 141.

40. See Pierre Gassendi, *The Selected Works of Pierre Gassendi*, trans. and ed. Craig B. Bush (New York: Johnson Reprint, 1972), 381. See also Margaret J. Osler, "Baptizing Epicurean Atomism: Pierre Gassendi on the Immortality of the Soul," in *Religion, Science, and Worldview: Essays in Honor of Richard S. Westfall*, ed. Margaret J. Osler and Paul L. Farber (Cambridge: Cambridge University Press, 1985), 163–83.

nation culminates in Mercutio's Mab—a denial that phantasms are substantial, visualized with a figural intensity that reflects the aesthetic excellence of *De rerum natura*. But to understand Mercutio's oration, we must situate him within the broader materialist concerns of the play, particularly the extended exploration of the relation between matter and mind articulated in its pervasive imagery of films, powders, and fumes. Viewed through this imagery, the stars, dreams, and plagues of *Romeo and Juliet* emerge as a chaotic material cosmology in which "substance" and "show" may be confusingly, sometimes thrillingly, elided. In the carnal fantasies of the eponymous lovers especially, even the slightest and flimsiest of things are suspected to possess a degree of material solidity, while the most solid-seeming forms belong to a mutable and particulate nether reality hidden from view.

Love in the world inhabited by Juliet and Romeo adheres in air, fire, vapor, and dust. It is weightless substance, matter attenuated to its thinnest extreme.

> A lover may bestride the gossamers
> That idles in the wanton summer air
> And yet not fall, so light is vanity. (2.5.18–20)

This marvelous lightness is noted by Friar Laurence, the play's toxicologist, who is able to alter conscious and corporeal states with the aid of fluids. He compares the delights of love to "fire and powder"—both destructively "consume" what they "kiss" (2.5.10–11). Desire may be light as air, but even air is governed by entropic laws. Echoing the Friar's logic, Benvolio believes that "one fire burns out another's burning" (1.2.46); by this reasoning, he says Romeo should stop pining for frosty Rosaline:

> Take thou some new infection to thy eye,
> And the rank poison of the old will die. (1.2.50–51)

Romeo himself conceives of love as a kind of exhalation,

> a smoke made with the fume of sighs;
> Being purged, a fire sparkling in lovers' eyes. (1.1.185–86)

He refers to the commonly held notion that humoral imbalances send up vapors into the brain, corrupting the imagination and producing erratic phantasms.[41]

---

41. See, for example, Levinus Lemnius, *The touchstone of complexions*, trans. Thomas Newton (London, 1576), P1r; William Perkins, *The whole treatise of the cases of conscience* (London, 1606), 192; and Sennert, *Thirteen books*, 386.

It is the lovers who are perhaps most intensely attuned to the ultrafine distinction between nothing and not-nothing. Thin substances are hard to make out: the progress of poison is concealed inside the body; fumes vanish; and webs are impossible to measure. Yet Romeo and Juliet attempt to take their measure all the same, with their imaginations. When he prompts her to describe their imminent "imagined happiness," she demurs, saying,

> Conceit, more rich in matter than in words,
> Brags of his substance, not of ornament. (2.5.30–31)

As Catherine Belsey notes, Juliet means that her love is beyond expression; "its substance cannot be counted, cannot be summed up in words."[42] But she may be using "conceit" in a cognitive sense too. The word was a name for imagination: Philip Sidney speaks in his *Defence of Poesy* of the *"idea* or foreconceit" of literary works.[43] Juliet's point is that verbal "ornament" is unequal to the richness of feeling—but what she is literally saying is that a fantasy weighs more than speech. Doomed and short-lived, Romeo and Juliet must live on the edge between the substantial and the insubstantial, the precarious place of the barely there.

In a similar way, thin "substance" connotes emotional intensity for Romeo. Light-headed with euphoria in the balcony scene, he wonders if

> all this is but a dream,
> Too flattering-sweet to be substantial. (2.1.182–83)

He uses the word again when he struggles to understand dead Juliet's lifelike appearance: "unsubstantial death is amorous," and he therefore keeps "Thee here in dark to be his paramour" (5.3.103–5)." Death is "unsubstantial" because he is skeletally "lean," and also because he keeps his concubine, Juliet, suspended in a materially ambiguous condition, vividly dead. At once, the lovers feel themselves connected to the very large and the very small. When Romeo resolves to die, he is relieved to "shake the yoke of inauspicious stars / From this world-wearied flesh" (5.3.111–12), a remark that echoes Juliet's earlier desire to see him cut "out in little stars" and strewn across the sky (3.2.22). Whereas she would make him into a constellation, he makes himself tiny: dismayed by his banishment, Romeo observes,

42. Catherine Belsey, "The Name of the Rose in *Romeo and Juliet*," *Yearbook of English Studies* 23 (1993): 138.

43. Philip Sidney, "The Defence of Poesy" in *The Major Works*, ed. Katherine Duncan-Jones (Oxford: Oxford University Press, 2009), 216.

More honorable state, more courtship, lives
In carrion flies than Romeo. They may seize
On the white wonder of dear Juliet's hand,
And steal immortal blessing from her lips. (3.3.34–37)

Lover and beloved are macabrely triangulated with a pest. Romeo envies the
fly that may crawl undetected on Juliet's hand and lip—a rather ghoulish pre-
vision of her death. Though he feels anguish, Romeo seems as well to be darkly
amused by his own fancy, punning, "Flies may do this, but I from this must
fly" (3.3.43).

It is possible to read these intimations of mortality as reminders about the
impermanence of life, consonant with the *memento mori* and *vanitas* traditions
of Renaissance art. As readers of this play commonly observe, it is filled with
poisons, corpses, and plagues.[44] Life in *Romeo and Juliet* is a rose ever on the
brink of extinction. A lover is a "bud bit with an envious worm" (1.1.146); "full
soon the canker death eats up that plant" (2.2.30). While these frequent reminders of death express the brevity of mortal life in a wholly conventional
manner, the play seems also to reach at a larger cosmological reality that extends beyond human existence.[45] Both Juliet and Romeo are prescient, able
to visualize the materiality of their bodies with startling clarity. Hearing of
Romeo's banishment, for example, Juliet instantly offers to entomb herself:

Hide me nightly in a charnel house,
O'ercovered quite with dead men's rattling bones,
With reeky shanks and yellow chapless skulls;
Or bid me go into a new-made grave,
And hide me with a dead man in his shroud. (4.1.81–85)

Juliet imagines acts of self-slaughter: a leap from a tower's "battlements,"
deadly "serpents," "roaring bears." In the final scenario, she does not so much

---

44. As a recent instance, see Mary Floyd-Wilson, "'Angry Mab with Blisters Plague': The Pre-
Modern Science of Contagion in *Romeo and Juliet*," in *The Palgrave Handbook of Early Modern Literature
and Science*, eds. Howard Marchitello and Evelyn Tribble (London: Palgrave Macmillan, 2017), 401–422.
See as well David Bergeron, "Sickness in *Romeo and Juliet*," *College Language Association Journal* 20
(1977): 356–64; Tanya Pollard, "'A Thing like Death': Potions and Poisons in *Romeo and Juliet* and *Antony and Cleopatra*," *Renaissance Drama* 32 (2003): 95–121; and Lynette Hunter, "Cankers in *Romeo and
Juliet*: Sixteenth-Century Medicine at a Figural/Literal Cusp," in *Disease, Diagnosis, and Cure on the
Early Modern Stage*, ed. Stephanie Moss and Kaara L. Peterson (Hants, England: Ashgate, 2004),
171–85.

45. Ramie Targoff has argued that Shakespeare amended his sources to deny Romeo and Juliet
reunion in an imagined life; see *Posthumous Love: Eros and the Afterlife in Renaissance England* (Chicago:
University of Chicago Press, 2014), chap. 4, 97–134.

die as become corporeally integrated with dead things, interred with the dead man in his "new-made grave." Darkly, the speech predicts the play's last act, in which Juliet is laid alongside dead Tybalt in the Capulet vault. More proleptic seeing follows when the Friar tells Juliet what to expect from the sleep-inducing poison:

> Through all thy veins shall run
> A cold and drowsy humor—for no pulse
> Shall keep his native progress but surcease. (4.1.95–97)

Immediately, Juliet begins to feel these things. The undrunk vial still in her hand, she senses

> a faint cold fear thrill[ing] through my veins,
> That almost freezes up the heat of life. (4.3.15–16)

A short while later, Juliet has a third presentiment: she imagines awakening without Romeo,

> stifled in the vault,
> To whose foul mouth no healthsome air breathes in. (4.3.33–34)

Again, she summons carnal feelings purely by the power of her imagination, inhaling the "loathsome smells" of that noxious place (4.3.46). She can even anticipate future fancies, the "horrible conceit of death and night" that would compel her to "madly play with my forefathers' joints" or cudgel her brains "with some great kinsman's bone" (4.3.51–53). In what Matthew Spellberg calls a perceptual loop of "touching-imagining-touching," her mind, touched by fancy, craves even more intense tactile experiences.[46] Romeo is not so different: when he arrives in the tomb to find Juliet dead, he enters into a decompositional fantasy of his own, lying down to await the "worms that are [her] chambermaids" (5.3.109).

These and other grisly images may be seen as attempts to expose the illusion of material form. The world of the play is propelled by disastrous conjunctions: young love is orbited by ancient hatred; life and death are locked in a tight embrace. Though these dialectical forces might seem inexorable, the

---

46. Matthew Spellberg, "Feeling Dreams in *Romeo and Juliet*," *English Literary Renaissance* 43, no. 1 (2013): 74.

lovers intuit that the substantive reality of their universe is likely different from the way it superficially appears. When, in the opening scene, Romeo finds evidence of Verona's latest brawl, he conceives an ontological chaos:

Here's much to do with hate, but more with love.
Why, then, O brawling love, O loving hate,
O anything of nothing first created,
O heavy lightness, serious vanity,
Misshapen chaos of well-seeming forms,
Feather of lead, bright smoke, cold fire, sick health,
Still-waking sleep that is not what it is—
This love feel I, that feel no love in this. (1.1.170–77)

Keening apostrophes string together a list of oxymorons: "brawling love," "loving hate," "serious vanity." Together, they figure civic strife as semantic rupture, as meaning unhinged. In one sense, the paradoxes intimate hypocrisy. Bellicose pride, says Romeo, serves as a mask for brutality; beneath "well-seeming forms" of nobility lies savagery. In a more materialist and perhaps existential sense, the violence itself is unreal, for below the "well-seeming forms" of the human world lies the basic "misshapen chaos" of matter, which is governed only by physical parameters such as weight ("lightness"), texture ("feather"), illumination ("smoke"), or heat ("fire"). On this plane there are no Capulets, no Montagues. Romeo's metaphysics is Christian in that it implicitly maintains that all things were first created "of nothing." Yet his lines also suggest that our notions of "anything" and "nothing" may be rather limited. The implications of this are so great that Romeo struggles to express them: formally, his speech is a chain of appositives that the listener must corral into meaning, much as the mind interprets matter as form. To sense what cannot be sensed, to grasp the existence of absence, is for Romeo a kind of sensation in itself: "This love feel I, that feel no love in this."

Juliet has a version of Romeo's unusual apperception. When she hears of the skirmish that has resulted in Romeo's slaying Tybalt, she too catches sight of a cosmic fissure:

O serpent heart, hid with a flow'ring face!
Did ever dragon keep so fair a cave?
Beautiful tyrant, fiend angelical,
Dove-feathered raven, wolvish-ravening lamb,
Despisèd substance of divinest show,

Just opposite to what thou justly seem'st,
A damnèd saint, an honourable villain. (3.2.73–9)

In a rare moment, Juliet's love for Romeo is shaken. Alarmed by the possibility that he may be other than what she imagined, she utters a string of ontological contradictions—"beautiful tyrant," "dove-feathered raven." Again, a cumulative mass of such phrases seems necessary to express the magnitude of her discovery. The notion that Romeo is an unknowable other can only be made graspable by pushing him through various metamorphoses: she makes him a serpent, a fiend, a bird, a wolf, a saint. Tropes cluster in Juliet's mind as she confronts the divorce between "despisèd substance" and "divinest show"; this realization compels her to reconceptualize the apparent divisions among all things. Caught as they are in an inhospitable cosmos, Juliet and Romeo are separately alert to the contrary realities of the world and predisposed to deconstruct the axiomatic dichotomy of "substance" and "show."

Normally, the familiar rhetorical pairing of "show" (or "shadow") with "substance" underscores the poverty of the former, as we saw in discussions of Christ's real presence in the sacrament of the altar.[47] In fact, the discrepancy between substance and accident was a problem for natural philosophers too, as we see in a moment of intellectual frustration in Sennert. Puzzling over the nature of light, he quotes Fracastoro on the pitiless "obscurity of Nature":

> When of late I busily Sought, those thinly appearing
> Images that flow from things and shed them all over,
> And to the Wayles woods went alone to the Silent
> Secret shades, there I found, how by these shews I deluded
> Was, though they smite always our sense and still they assail us.[48]

Wishing to see the "thinly appearing" atomic simulacra—the "images that flow from things"—the philosopher retires to "Wayles woods" and "secret shades." But there he finds nothing other than his own delusive mind. The corpuscular "shews," though they are constantly entering and exiting "through the Doors of our senses," remain stubbornly undetectable. In *Romeo and Juliet*, the supposed contrast between substance and show is interestingly muted. Instead, the implication is that substantiality itself is a deceptive fiction, for even the slightest and flimsiest of shows contain matter, while the apparently weighti-

---

47. See, for example, John Jewel, *A replie vnto M. Hardinges answeare* (London, 1565), 147, 421, 435, 455, 462. See also Stuart Clark, *Vanities of the Eye: Vision in Early Modern European Culture* (Oxford: Oxford University Press, 2007), 187.

48. Sennert, *Thirteen books*, 57.

est ones are perforated in a way that our senses cannot perceive. As though sensitized to this paradox, Juliet and Romeo test traditional notions of materiality; they are cosmologists in their way.[49]

Thus, imagination in *Romeo and Juliet* is a carnal or material experience as much as a spiritual or mental one. For its young protagonists, reality is more interesting when it carries the dreamlike texture of imagination, when it provokes the suspicion that there is a rift between that which is sensed and that which is. The desire is not so much to pull back the flesh in order to reveal the grinning skull below as to take the cover off the entire semblance of material form itself, to apprehend instead the fine, filmy, and fiery substances that lie beneath. Phantasms too are subtly incorporated in the entropic economy that governs everyday phenomena such as combustion, decomposition, or osmosis. The lovers' fantasies evade categorization; part dream, part contagion, and part prophecy, their phantasmatic intuitions suggest imagination's still unclear but unmistakably real place in the larger cosmos. The force of Shakespeare's tragedy is deepened by its implicit idea that the finely textured, life-devouring love of Romeo and Juliet is no more than an epiphenomenon, an illusory effect produced by an in fact indifferent, insensible cosmos. Their passion is a grand—the grandest—mirage of the material world; sensing this, their instinct is to examine the illusion all the more intently.

Vain Fantasy

This frontier between air and mind fascinates Mercutio as well, perhaps more than he knows. He too is embedded in *Romeo and Juliet*'s particulate universe of air and insects: he thinks of Verona's fashionmongers as "affecting fantasies" and "strange flies" (2.3.26–30). But whereas Romeo and Juliet trust in the tactility of their imaginations, Mercutio is skeptical, as is made clear in the Queen Mab sequence. Critics have not always found it easy to integrate Mercutio's rant about fairy dreams, which seems more apropos of *A Midsummer Night's Dream*, into the rest of Shakespeare's tragedy.[50] When read with Epicurean atomism and *De rerum natura* in mind, Mercutio's flight of fancy appears more obviously to participate in the play's thematic focus on the tangibility of thoughts. Indeed, it goes further, making a metaliterary statement about the scientific function of figurative language. Mercutio's attempt

---

49. For more on the play's cosmological imagery, see Mark Stavig, *The Forms of Things Unknown: Renaissance Metaphor in "Romeo and Juliet" and "A Midsummer Night's Dream"* (Pittsburgh: Duquesne University Press, 1995).

50. See, for example, Stanley Wells, *Shakespeare, Sex, & Love* (Oxford: Oxford University Press, 2010), 155.

to disprove dreams—a disproof that itself takes an unforgettably fantastical rhetorical form—gets away from him, in a way that suggests the extent to which natural philosophy makes use of sense-based imagery and depends heavily on gross imagination. In order to see this, it is necessary to contextualize the Mab excursus within the wider dramatic arc of the scene.

The scene begins with an epistemological dispute that primes us to be thinking about the relation between symbolic representation and sensible experience. As the Montague men make their way to the Capulet ball under the cover of night, Mercutio chides Romeo for "burning daylight"—dragging his feet. Romeo protests that he is doing no such thing, for it is already dark. Mercutio tells him not to be obtuse: what he means is that "We waste our lights in vain, light lights by day" (1.4.43). "Take our good meaning," he adds,

> for our judgment sits
> Five times in that ere once in our fine wits. (1.4.44–45)

For Mercutio, good "judgment" trumps information gleaned by lesser "wits"; it is "judgment" that enables the mind to parse complex figurative "meaning"—to work out, in this case, that lost light means a wastage of time. But Romeo is a literalist; his standpoint is an Epicurean one. Epicurus had maintained that impressions garnered by the senses are beyond question: as the axiom runs, "If you quarrel with all your sense-perceptions, you will have nothing to refer to in judging even those sense-perceptions which you claim are false."[51] The fact that a stick partly submerged in water appears broken is not the fault of the eye but of the mind, which has made an incorrect deduction. The "judgment" that Mercutio privileges is, according to this perspective, responsible for delusion. His position is understandable. As Plutarch points out in his discussion of Epicurus, to be told that one must believe one's senses completely makes living in the often illusionistic world a fearsome prospect, for it invites much "perturbation and ignorance about sensible things and imaginations."[52] Shakespeare has thus set up this scene so as to broach the difficulty of reconciling perception, judgment, and language at the outset.

Then the dialogue turns toward dreams. As though to explain his misgivings, misgivings that stem from imaginative intuition rather than rational judgment, Romeo begins, "I dreamed a dream tonight" (1.4.48). Mercutio dismisses this, interjecting that he too has dreamed—dreamed, in fact, "that

51. Epicurus, *The Epicurus Reader: Selected Writings and Testimonia*, trans. and ed. Brad Inwood and L. P. Gerson (Indianapolis: Hackett, 1994), 34. See also Andree Hahmann, "Epicurus on Truth and Phantasia," *Ancient Philosophy* 35, no. 1 (2015): 155–82, esp. 170–72.
52. Plutarch, *Morals*, 1125.

dreamers often lie" (1.4.49). Romeo persists, countering that dreamers "dream things true" (1.4.50).[53] On this question too, the men are divided. Like the ancients and medievals, the early moderns distinguished between different sorts of dreams.[54] Thomas Hill's treatise on dream interpretation, for example, speaks of "true dreames," which "foreshewe matters imminent," and the more "vain" kind that result from humoral effluvia.[55] We are not told what Romeo dreamed, only that he thinks dreams are "true." Judging from the dream he later mentions in act 5, in which "my lady came and found me dead" (5.1.6), it does seem that Romeo's imagination, like Juliet's, is prophetic.[56] A bit oddly, though, Romeo takes this to "presage some joyful news." For him, mental visions are felicitous solely by virtue of their nature.

Perhaps knowing this, Mercutio launches into a lengthy refutation of dreams. He does this by portraying Queen Mab, an impossibly tiny fairy, in intricate detail:

> She is the fairies' midwife, and she comes
> In shape no bigger than an agate stone
> On the forefinger of an alderman,
> Drawn with a team of little atomi
> Over men's noses as they lie asleep,
> Her wagon-spokes made of long spinners' legs,
> The cover of the wings of grasshoppers,
> Her traces of the smallest spider web,
> Her collars of the moonshine's wat'ry beams,
> Her whip of cricket's bone, the lash of film,
> Her wagoner a small gray-coated gnat
> Not half so big as a round little worm
> Pricked from the lazy finger of a maid.
> Her chariot is an empty hazelnut
> Made by the joiner squirrel or old grub—
> Time out o'mind the fairies' coach-makers. (1.4.52–67)

---

53. Shakespeare's dialogue is reminiscent of Timothy Bright, *A treatise of melancholie* (London, 1586), 119: "You will say such dreames are oft times but fancies. True: and many times they be no fancies."

54. For an overview of early modern dream theory, see Marjorie Garber, *Dream in Shakespeare: From Metaphor to Metamorphosis* (New Haven, CT: Yale University Press, 1974), 1–14.

55. Thomas Hill, *The most pleasaunte Arte of the interpretacion of Dreames* (London, 1576), D2v–D3r.

56. Dreaming of oneself dead, confirms Hill, is a harbinger of "hinderance." Juliet's dreams of carcasses and dead men's bones, meanwhile, suggest "straungenes and troubles." See Hill, F4r–F4v, P1r, P3v.

The portrait is specially designed to draw our attention to tiny objects and delicate substances. It focuses on the microscopic organs of insects—spiders' legs, crickets' bones—as well as things we would have trouble seeing and feeling—"web," "beams." The speech flaunts its artifice; itself a verbal artifact, it is moreover filled with artifacts: the wagon spokes "made of" spiders' legs; the wagoner's tiny "coat"; the "coach-makers'" chariot. Mercutio reduces and enlarges the frame of reference as he goes, moving from "squirrel" and "old grub" down to the imperceptible "atomi," drawing us from the visible to the subvisible.[57] Lucretius does something similar: "there are some living creatures so small that their third part cannot possibly be seen. What must you suppose one of their guts is like? the ball of the heart, or the eyes? the limbs and members? How small are they? What further of the first-beginnings which must compose the nature of their mind and spirit? *Do you not see* how fine and how minute they are?" (4.116–22, my emphasis). Through a series of gradations, Lucretius brings us to perceive the infinitesimal scale of "first-beginnings." Mercutio's lens operates somewhat more erratically, zooming in and out, tumbling from "grasshoppers" to "worm," from "wing" to "web." The shifting assemblage of parts, creatures, and textures self-consciously asserts the fictiveness of its scene, its status as an imagined representation, in a way that Lucretius does not.

After detailing Mab's entourage, Mercutio describes her nightly occupation:

And in this state she gallops night by night
Through lovers' brains, and then they dream of love;
On courtiers' knees, that dream on curtsies straight;
O'er lawyers' fingers, who straight dream on fees;
O'er ladies' lips, who straight on kisses dream,
Which oft the angry Mab with blisters plagues
Because their breaths with sweetmeats tainted are.
Sometime she gallops o'er a courtier's nose,
And then dreams he of smelling out a suit;
And sometime comes she with a tithe-pig's tail,
Tickling a parson's nose as 'a lies asleep—
Then he dreams of another benefice;
Sometime she driveth o'er a soldier's neck,
And then dreams he of cutting foreign throats,

---

57. Marie Garnier-Giamarchi also suggests that the disruptive poetics of Mercutio's speech mimic the atomic swerve. See "Mobility and the Method: From Shakespeare's Treatise on Mab to Descartes' *Treatise on Man*," in *Textures of Renaissance Knowledge*, ed. Philippa Berry and Margaret Tudeau-Clayton (Manchester: Manchester University Press, 2003), 139.

Of breaches, ambuscadoes, Spanish blades,
Of healths five fathom deep—and then anon
Drums in his ear, at which he starts and wakes
And, being thus frighted, swears a prayer or two
And sleeps again. (1.4.68–86)

The lady dreams of lovemaking, the soldier of carnage, the parson of riches. As Shakespeare's audience would have understood, the dreams engendered by Mab are those of the grossest sort, unsaintly fancies of the flesh. Dreams, Mercutio is saying, are little more than the carnal wishes of waking life luridly repackaged in sleep. No revelations are to be found here, unless it be thought very profound that courtiers dream of curtsies and lawyers of fees. More surprising is the implication that dreams are transmitted by touch. Unimpeded by physical barriers, Mab crawls on the outward contours of the body—"on courtiers' knees," "o'er lawyers' fingers," "o'er ladies' lips." She blisters mouths, tickles noses, and drums in ears, bothering the sense organs that were the supposed points of entry for Epicurean *eidola*. The picture is doubly uncomfortable. For one thing, it points out that, in a particle-based world, much life teems below the threshold of perception. Plutarch tried to alleviate this discomfort, stressing that the presence of atoms need not threaten our peace of mind: "In all this there is nothing neere at hand *to touch us*," says the English translation, "rather every one of these questions [is] farre remote, and beyond our senses."[58] The portrait is all the more discomfiting because it suggests that the teeming happens within our heads, in the throes of the unruly imagination.

This unruliness rears up especially in the last and darkest movement of the speech, when Mercutio seems to succumb to precisely the kind of imaginative excess he is cautioning against:

This is that very Mab
That plaits the manes of horses in the night,
And bakes the elflocks in foul sluttish hairs
Which, once untangled, much misfortune bodes.
This is the hag, when maids lie on their backs,
That presses them, and learns them first to bear,
Making them women of good carriage.
This is she—(1.4.86–93)

---

58. Plutarch, *Morals*, 1125, my emphasis.

We have entered into "a kind of possession."[59] Mab is now a "hag," a puckish sprite that troubles horses and seduces maids. Her movements resemble those of the devil, who "creepeth throughout all the passages of the senses" and "filleth all the passages of the intelligence with certayne mistes and clowdes."[60] As Mercutio's imagination wanders, the dream turns to nightmare, endlessly generating itself; Mab the "midwife" delivers more and more mental offspring. In turn, the three sections of Mercutio's speech—the miniature carriage with its miniscule parts, the creeping invasion of sleepers' bodies, and the brimming frenzy at the end—have presented the fancy as particle, intrusion, and pullulation.

"Thou talk'st of nothing," Romeo interrupts. And, wrested from his dream, Mercutio readily agrees, calling dreams

> the children of an idle brain,
> Begot of nothing but vain fantasy,
> Which is as thin of substance as the air
> And more inconstant than the wind. (1.4.95–98)

Having mimetically evoked the material imagination so convincingly that he practically falls under its spell, Mercutio abruptly consigns "vain fantasy" to thin "air." The coda is a little glib, not least because Mercutio has just demonstrated the extent of imagination's hypnotizing power. It seems ironic too, given that, for early scientists, "air" was certainly not nothing. We saw that air and imagination are linked in *A Midsummer Night's Dream*, where Theseus says that "imagination bodies forth / The forms of things unknown," that the poet's pen lends shape to "airy nothing" (5.1.14–16). There, the unusual use of *body* as a verb underlines the materiality of phantasms in a way different from the more abstract "shape," the more nebulous "thing," or the Platonic "form." In both speeches, a tremor of ambiguity attaches to the intuitive association of airiness and nothingness. Airborne disease provides the terrifying climax of Lucretius's poem, in fact; it ends with apocalyptic scenes of the Athenian plague, with pestilence falling from the sky in clouds and mists, and oozing out of the damp earth (6.1098–102). Air can penetrate the body even to the core, Sennert would write: the liquid metal that shares Mercutio's name was used to fumigate syphilis patients, a process that deposited "smal Atomes"

59. Joseph A. Porter, *Shakespeare's Mercutio: His History and Drama* (Chapel Hill: University of North Carolina Press, 1988), 104.

60. Le Loyer, *Treatise of specters*, 123v. On Mab as a depiction of the supernatural causes of infection, see Floyd-Wilson, "'Angry Mab,'" esp. 402–3.

of quicksilver even in "the Veins and Bones."[61] Mercutio's reference to "wind" is notable also. Not only is wind a favorite Lucretian example of the large-scale havoc that atoms can wreak; it is also an analogue for the soul, imagined as an airy or smoky thing nestled in the body until death, when it disintegrates and disperses.[62] Finally, air is the medium of dreams, specters, and other apparitions, as Stuart Clark writes; it is "an element of sleights and tricks, not one of transparency and truth."[63]

The fact that Mercutio is oblivious to these ironies turns his oration into something more than just a materialist—or antimaterialist—rant. What he has presented is not exactly a philosophy of perception. He has avoided saying exactly what Mab does; phrases such as "and then they dream" and "then dreams he" imply correlation rather than causation. Then there is the strange characterization of Mab, which cannot be attributed to Lucretian philosophy. Demons and angels often feature in early modern oneiric theory; atoms and insects spill into representations of imagination. Fairies rarely come up in either.[64] Nor does Mab much resemble Shakespeare's other fairies—neither the lordly Oberon and Titania nor their obedient pages Moth and Cobweb.[65] Mab is perhaps best understood less as a particular fiction than as a symbol of figuration. The analogies on which atomist explanations are founded, Mary Thomas Crane points out, are structural comparisons, not occult resemblances of the kind described by magicians and alchemists.[66] They have no purpose beyond helping us conceptualize the material reality of the universe. Mab is a self-conscious statement about the essential role of metaphorical language in scientific thinking. She moreover demonstrates the double-edged power of metaphor. Contagion, wrote Fracastoro, depends on contact: there is a source, a target, and a carrying over from one to the other.[67] This is also how meta-

---

61. Daniel Sennert, *The institutions or fundamentals of the whole art, both of physick and chirurgery* (London, 1656), 228; Sennert, *Thirteen books*, 453.

62. See Lucretius, *Nature of Things*, 3.119–29; and Charles Segal, *Lucretius on Death and Anxiety: Poetry and Philosophy in "De Rerum Natura"* (Princeton, NJ: Princeton University Press, 1990), 46, 48, 66.

63. See Clark, *Vanities of the Eye*, 248: Le Loyer and "other participants in the apparitions debate" understood specters to be produced by "the reflective qualities of air," which acted like a "gigantic looking-glass."

64. That is, until the mid-seventeenth century, when Cavendish would repeatedly connect fairies and brains; see, for example, her *Poems, and fancies*, 162: "in the *Braine* may dwel / Little small *Fairies*" who "make / Those *formes* and *figures*, we for *fancy* take."

65. On possible folkloric sources for Mab, see Diane Purkiss, *At the Bottom of the Garden: A Dark History of Fairies, Hobgoblins, and Other Troublesome Things* (New York: New York University Press, 2001).

66. Crane, *Losing Touch with Nature*, 133.

67. Girolamo Fracastoro, *Hieronymi Fracastorii De contagione et contagiosis morbis et eorum curatione*, trans. Wilmer Cave Wright (New York: G. P. Putnam's Sons, 1930), 3.

phor works, of course—it is a rhetorical "Figure of Transport," as George Puttenham says, whereby a word is relocated from "his own right signification to
another not so natural."[68] Mercutio, who speaks easily of daylight where there
is none, who thinks in metaphors, invents Mab as his figure. But then
he becomes infected by her, has trouble shaking her off. Something similar
could be said about the importation of Lucretius into the discourse of imagination. With indisputable efficacy, atoms and insects depict the fantasy's
power to diffuse, multiply, and roam. Less favorably, they also bind the image-
making faculty to triviality, putrescence, and infestation.

Ironically, Mercutio, in his attempt to expound the nature of "nothing," has
had to invent a great many somethings. With this, Shakespeare exposes the
reliance of theoretical physics on insubstantial fantasies. Because the human
mind cannot perceive the substantive basis and material mechanisms of the
universe as they are, it falls to representational devices to translate imperceptible phenomena into comprehensible terms—and this, in effect, means appealing phantasmatically to the physical senses. Touch is the most concrete
of these; it happens also to be how atoms interact, the "ontological mechanism that glues reality together."[69] And so a theologian seeking to undo a fantastical heresy speaks of flying motes in consecrated bread, and a skeptical
trickster concocts a lavishly accoutered fairy to establish the impossibility of
her existence. The fact that metaphysical ideas invariably bear a palimpsest
of the physical world explains why "vain fantasy" is contradictorily imagined as both empty and full: we can only talk about voids by talking about
bodies. Tickling Mab thus shows that the human understanding of cosmological laws is filtered through the approximations of linguistic representation
and that language is in turn anchored in the wordless, sensing life of the
body.

In the end, Mercutio's performance does not erase in Romeo the foreboding
produced by his undisclosed dream; though Romeo may have been tricked
into saying that to talk of dreams is to talk of "nothing," he does not think his
fancies are the miscarriages of an idle brain, no more than does Juliet. Nonetheless, the scene intimates, as does the play as a whole, how the proverbial
vanity of imagination was complicated during the Renaissance, thanks to increasingly nuanced cultural and natural philosophical theories of matter. This
more complicated, more materially ambiguous "vain fantasy" could in turn

---

68. George Puttenham, *The Art of English Poesy*, ed. Frank Whigham and Wayne A. Rebhorn
(Ithaca, NY: Cornell University Press, 2007), 262–63.

69. Pablo Maurette, "De rerum textura: Lucretius, Fracastoro, and the Sense of Touch," *Sixteenth
Century Journal* 45, no. 2 (2014): 312, 315.

serve as a tool for conceptual and rhetorical discussions of substance, in deciding not only what is solid and holds weight but also what holds meaning,
is real or true. *Romeo and Juliet*, which holds in its sights both the microscopic
atomi and the planetary macrocosm, considers where on the cosmological
spectrum phantasms belong.

CHAPTER 4

Seeming to See

King Lear's Mental Optics

                    Though the early moderns analogized the imaginative faculty in many ways, no analogy could outdo the one that had been inscribed in the notion of *phantasia*—indeed, in the word itself—from its earliest beginnings: to imagine is to see inwardly. In faculty psychology, the eye and the mind's eye exist in a singular relation. None of the other physical senses had an analogue among the faculties, and none of the other faculties was compared to an organ of the body. For a literary illustration of this resemblance between natural and mental light, of its luminous aesthetic power, we can look to Shakespeare's sonnet 43:

> When most I wink, then do mine eyes best see,
> For all the day they view things unrespected,
> But when I sleep, in dreams they look on thee,
> And, darkly bright, are bright in dark directed.
> Then thou, whose shadow shadows doth make bright—
> How would thy shadow's form form happy show
> To the clear day, with thy much clearer light,
> When to unseeing eyes thy shade shines so!
> How would (I say) mine eyes be blessèd made
> By looking on thee in the living day,
> When in dead night thy fair imperfect shade

Through heavy sleep on sightless eyes doth stay!
  All days are nights to see till I see thee,
  And nights bright days when dreams do show thee me.[1]

The poem begins by reviewing the epistemological problems created by lovestruck eyes, the same sort of problems that we noted in chapter 1: they "view things unrespected," sensing but not heeding. Soon, though, the emphasis shifts to the beloved's phantasmatic "shade" and the way it shines in the lover's dreaming mind, illumining the dark "night" of separation. The chiastic repetitions that begin at the end of the first quatrain—"darkly bright, are bright in dark"; "shadow shadows"; "shadow's form form happy show"—are suggestive of mirror images. The thorough entangling of the words *dark*, *bright*, and *shadow* renders the mind's interior a spectral gloom, filled with intermelting light effects. Looking closely, we find that the visual echoes are inexact: "darkly" reappears as "dark"; "shadow" as "shadows," "shadow's," and the contracted "show"; even "form form" is imperfect, with the word morphing from noun into verb in its second iteration. With these several slanted reflections, the poem introduces a tension of difference within the similarity it ostensibly celebrates. Ultimately, both "imperfect" dream and "unrespected" day are flawed states. Ideal vision—that is to say, the "blessed," "clear day" of the beloved's return—remains beyond the sonnet's temporal horizon.[2]

The fascination that Shakespeare and other Renaissance poets had for the light of the mind was likely deepened by the way the connection between vision and mental representation was shifting at the time. While the metaphorical possibilities of atomic phantasms, say, were breeding, the age-old association of seeing and imagining, this chapter will suggest, was past its prime, in need of revision. For much of the history of visual theory, the sensory and the cognitive aspects of sight had remained closely allied, if not thoroughly contiguous. When, at the turn of the seventeenth century, it was demonstrated that the human eye works like a camera, however, optics became a matter primarily of mathematics; seeing was effectively decoupled from imagining. Nevertheless, in ophthalmological texts, Neoplatonist treatises, lens-making handbooks, and Renaissance poetry, there was a great deal of interest in the supposed optical properties of imagination. For Shakespeare,

---

1. Shakespeare references are from *The Norton Shakespeare*, ed. Stephen Greenblatt et al. (New York: W. W. Norton, 2016), hereafter cited parenthetically.

2. Other poets wrote sonnets on the same conceit: see, for example, Philip Sidney, *Astrophil and Stella* 38, in *The Major Works*, ed. Katherine Duncan-Jones (Oxford: Oxford University Press, 2009), 167: "This night while sleep begins with heavy wings"; and Barnabe Barnes, *Parthenophil and Parthenophe* 88 (London, 1593): "Within thine eyes mine hart takes all his rest."

the unmooring of fantasy and sight provided a means of thinking about universal upheaval. We see this in *Venus and Adonis*, where a moment of perceptual shock initiates an era of perpetual dysfunction. We see it as well, more cataclysmically, in the upside-down world of *King Lear*, where the elegant conceptual symmetries between vision and imagination are themselves upended. Shakespeare considers the extent to which the image-making mind can correct failures of perception. Against the backdrop of early modern optics, his tragedy imagines a paradigmatic shift of its own, one that exchanges the unrealizable ideal of acuity for an endless sequence of distortions and corrections.

FANCY'S GLASS

Around the time that Shakespeare lived, the study of optics dissolved its interest in mental perception, and the discourse of imagination intensified its interest in optical phenomena. To see how this came about, it is necessary to return to antiquity, when the study of light was jointly philosophical, medical, and mathematical in nature and theories of vision presumed a connection between perceiver and perceived.[3] Plato held that vision occurs when the eyes emit a fiery beam that coalesces with natural light, welding "like unto like" to produce a "kindred substance" that conveys the sensible properties of the visual object to the observer's soul.[4] Aristotle, in contrast, said that light enters rather than exits the eye. Yet he too posited that the "sentient subject" becomes "like" its object, "shares its quality"; vision in his system depends on the fact that eye and air share in the quality of transparency.[5] Galen, in conjunction with his work on ocular anatomy, developed the idea that the brain sends a luminous *pneuma* out through the eye, with which it fetches information about the outside world.[6] These accounts proceed from an assumption that object, eye, and mind are fundamentally enchained in the visual process.

The branch of optics that would win out in the early modern period did not do this. On the premise that light is propagated along rectilinear rays, Euclid's *Optica* treated vision as a set of geometric problems, illustrating with

---

3. For historical overviews, see David C. Lindberg, *Theories of Vision from Al-Kindi to Kepler* (Chicago: University of Chicago Press, 1976); Olivier Darrigol, *A History of Optics: From Greek Antiquity to the Nineteenth Century* (Oxford: Oxford University Press, 2012); and A. Mark Smith, *From Sight to Light: The Passage from Ancient to Modern Optics* (Chicago: University of Chicago Press, 2015).

4. Plato, *Timaeus*, in *Timaeus. Critias. Cleitophon. Menexenus. Epistles*, trans. R. G. Bury, Loeb Classical Library 234 (Cambridge, MA: Harvard University Press, 1929), 45c.

5. In Aristotle, *On the Soul. Parva Naturalia. On Breath*, trans. W. S. Hett, Loeb Classical Library 288 (Cambridge, MA: Harvard University Press, 1957), see *On the Soul*, 418a1–6, and *On Sense and Sensible Objects*, 438a13–17.

6. Rudolph E. Siegel, *Galen on Sense Perception* (Basel, Switzerland: Karger, 1970), 58.

diagrams why, for example, objects at a distance appear to be smaller and how the height of a tall structure may be calculated by measuring its shadow.[7] Through subsequent works by Ptolemy, Alhazen, and others, mathematical optics was handed down to the fifteenth century, where it would find novel applications in art and architecture: Filippo Brunelleschi invented the use of linear perspective in painting, principles of which Leon Battista Alberti would publish in his *Della pictura* of 1435. Whereas medieval scholars had taken pains to reconcile the mathematical tradition with philosophical and anatomical ones, Renaissance polymaths were not as concerned with matters of sensation and perception, Martin Kemp notes: "If the eye figures little, the mind features even less."[8] Mathematics now offered a way of understanding visual illusion without recourse to cognitive theory. As the sixteenth-century astrologer John Dee wrote in his preface to Euclid's *Elements of geometrie,* "perspective" reveals the reasons behind marvels such as rainbows, comets, and the iridescent tails of peacocks and so elevates the dignity of humankind. Rather than be "ouershot and abused" by our eyes, he says, we will in the future use them "with greater pleasure: and perfecter Iudgement."[9]

By the end of the century, the cognitive theory of phantasms—at whose etymological root, of course, lay *phaos,* the Greek word for light—was mostly shut out of the study of optics. A landmark moment came in 1604, when the German astronomer Johannes Kepler showed in his *Ad Vitellionem paralipomena* that the visual image is formed on the retina, the photosensitive membrane at the back of the eye. Light rays enter and are refracted through the crystalline lens, projecting a two-dimensional picture, smaller in scale and oriented upside down, on the eye's far side (see figure 5). In effect, Kepler had demonstrated that the eye is a camera obscura, a closed box with an aperture covered by a lens. He was not the first to think this: Leonardo da Vinci had arrived at a similar notion a century earlier, realizing too that the image would be inverted and speculating as to how it was set right side up.[10] The natural philosopher Giambattista della Porta had likewise suggested in 1558 that the pupil is akin to "the hole of a window" in "a dark Chamber."[11] Experimenting with pinhole cameras around the same time were Girolamo Cardano and Daniele Barbaro, the latter of whom was a translator of Vitruvius's

7. Euclid, "The Optics of Euclid," trans. Harry Edwin Burton, *Journal of the Optical Society of America* 35, no. 5 (1945): 357–72.

8. Martin Kemp, *The Science of Art: Optical Themes in Western Art from Brunelleschi to Seurate* (New Haven, CT: Yale University Press, 1990), 165.

9. John Dee, preface to Euclid, *The elements of geometrie,* trans. Henry Billingsley (London, 1570), b1r–b1v.

10. Lindberg, *Theories of Vision,* 164, 166.

11. Giambattista della Porta, *Natural magick* (London, 1658), 364–65.

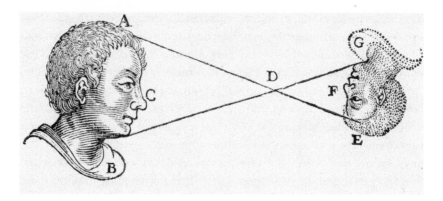

FIGURE 5. Diagram illustrating how an image ACB is inverted as EFG when light rays pass through a lens at aperture D. From René Descartes, *La dioptrique* (Leiden, 1637), 47. RB 336029, Huntington Library, San Marino, California.

De architectura.[12] Separately, the physician Felix Plater had surmised that the retina rather than the eye's center was the "sensitive organ of vision."[13]

Kepler's achievement was to bring all this together, culminating the work of his early modern and medieval precursors. The *Paralipomena* had important consequences: it made clear that vision does not happen in the eye's watery interior, as was previously thought, and that sight does not involve extramission. Most significantly, perhaps, it established that the visual image is a picture—Kepler introduced the word *pictura* into optical analysis—that is completely independent of the observer.[14] The remarkable implication, Svetlana Alpers writes, is that seeing does not rely on human intervention: the eye is "a dead eye," merely a physical site on which the world is constantly "picturing itself in light and color."[15] As René Descartes would later observe in his treatise on optics, it was no longer necessary to understand how mental "images themselves resemble their objects," for there was no such resemblance.[16]

That being said, certain points remained unresolved. Kepler was silent, for instance, on the problem of how the retinal inversion is corrected. It was a

12. Girolamo Cardano, *The "De Subtilitate" of Girolamo Cardano*, ed. John M. Forrester (Tempe: Arizona Center for Medieval and Renaissance Studies, 2013), 1:250.

13. A. C. Crombie, "Early Concepts of the Senses and the Mind," *Scientific American* 210 (1964): 110.

14. Lindberg, *Theories of Vision*, 202.

15. Svetlana Alpers, *The Art of Describing: Dutch Art in the Seventeenth Century* (London: Murray, 1983), 36.

16. René Descartes, *Discourse of Method, Optics, Geometry, and Meteorology*, trans. Paul J. Olscamp (Indianapolis: Hackett, 2001), 90.

quest to rectify errors that had originally driven him to study optics: while trying to resolve astronomical anomalies noted by Tycho Brahe, it had occurred to Kepler that the fault lay with his eyes, not his equipment. Having come to understand those flaws, Kepler was content to accept them. In his view, the eye was not diminished as an epistemological instrument by its idiosyncrasies; the mathematician chose simply to "take the measure of its deception or artifice."[17] The other point Kepler did not explain is what happens to the retinal image after its formation: "How the image or picture is composed by the visual spirits that reside in the retina and the nerve, and whether it is made to appear before the soul or the tribunal of the visual faculty by a spirit within the hollows of the brain, or whether the visual faculty, like a magistrate sent by the soul, goes forth from the administrative chamber of the brain into the optic nerve and retina to meet this image, as though descending to a lower court—this I leave to be disputed by the physicists."[18] Kepler admits his ignorance of the psychological and anatomical aspects of perception (denoted here by terms such as "spirits," "visual faculty," and "optic nerve"). Like Vesalius, he refuses to speculate about mental processes. Kepler goes further, however—he narrows the disciplinary bounds of optics so as to exclude questions of cognition altogether. He asks, rhetorically, "what can be said, according to the laws of optics, concerning this hidden journey, which, since it occurs through opaque and therefore dark places and is brought about by means of spirits which differ totally from humors and other transparent things, already exempts itself entirely from optical laws."[19] The tail end of the perceptual process is a "hidden journey" through "opaque and therefore dark places" (opacas, ideoque tenebrosas partes), driven by spirits that are not "transparent" and therefore excepted from the laws of light. Ironically, in order to make this point, the author adopts the language of light speaking of darkness and opacity in those spaces of the body where—as he is right now arguing—they cannot strictly be considered to exist.

The Paralipomena's prefatory materials similarly underline the intimate bonds between eye and mind in a manner somewhat contrary to the discovery of the main text. A first epigram triumphantly beckons as readers all those who are interested to see the "fiery eye." Let them not, it says, fear "flying

---

17. Alpers, Art of Describing, 35.

18. This is Lindberg's translation in Theories of Vision, 203. For the original, see Johannes Kepler, Ad Vitellionem paralipomena quibus astronomiae pars optica traditur, ed. Franz Hammer, in Gesammelte Werke, ed. Walther von Dyck and Max Caspar (Munich: C. H. Beck, 1938), 2:151–52.

19. Lindberg, Theories of Vision, 204; Kepler, Paralipomena, 152.

atoms," varicolored specters, or compound mirrors.[20] Cleverly alluding to various topics in the history of optics—extramission, the particulate *eidola* of the Epicureans that we noted in the previous chapter, catoptrics—it wittily promises that the present volume is going to deliver "things never before seen by the eye" (*oculis nunquam visa*). These insights it majestically attributes to the brilliant eyes of the author's mind (*oculis animi*)—Kepler's own imagination. There happens to be a second, "Author's Epigram" too, written as a decorous dialogue between Eyes and Mind. The former, saddened to have lost their sharpness, proffer their light to the latter, pleading, "deliver us from darkness" (*eripe nos tenebris*). Mind graciously assents, vowing to shine brightly despite its own imperfections (*naeuis irradiabo iubar*). The mathematics be as they may, the seamless harmony long imagined to exist between corporeal and mental vision was not about to vanish.

This may have been in part because another channel in the history of optics had helped consolidate the notion that imagination too works like a kind of optical instrument. One of the reasons that optics had thrived throughout the Middle Ages was its high theological stakes. Because light, as Dee says in his preface to Euclid, was "the first of *Gods Creatures*," its study promised "some kind of rationalization of how God's grace pervaded the universe."[21] The notion of archetypal divine illumination was no metaphor—in the cosmogony of Robert Grosseteste, for example, metaphysical *lux* is directly responsible for physical *lumen*.[22] Nor were Christian optics incompatible with Greek and Arabic philosophy, for medieval scholars had deftly synthesized all three. In terms of human psychology, it was thought that the soul "needed enlightenment from some other light source in order to participate in the truth," as Augustine put it.[23] To explain the soul's interaction with God, there was an oft-quoted analogy involving cognition and eyesight, reiterated by Marsilio Ficino as follows: "Just as the eyesight nowhere sees anything visible except in the radiance of the highest thing visible, that is, the Sun itself, so the human intellect apprehends nothing intelligible, except in the light of the highest thing

---

20. This epigram is credited to Jan Seuss, secretary to Christian II of Denmark. It and the next one may be found in Kepler, *Paralipomena*, 13. See as well William H. Donahue's translations, Johannes Kepler, *Optics* (Santa Fe: Green Lion, 2000), 11.

21. See, respectively, Dee, *Elements*, b1r; Samuel Y. Edgerton Jr., *The Renaissance Rediscovery of Linear Perspective* (New York: Harper and Row, 1976), 60.

22. Christina Van Dyke provides an overview of Grosseteste's theological optics in her essay "An Aristotelian Theory of Divine Illumination: Robert Grosseteste's *Commentary on the Posterior Analytics*," *British Journal for the History of Philosophy* 17, no. 4 (2009): 685–704.

23. Augustine, *Confessions, Volume I: Books 1–8*, trans. Carolyn J.-B. Hammond, Loeb Classical Library 26 (Cambridge, MA: Harvard University Press, 2014), 175.

intelligible, that is, in the light of God."[24] The precise way in which this apprehension happens was increasingly figured with the aid of mirror imagery. A comment by Plato saying that the liver is like "a mirror which receives impressions" from "the power of thoughts which proceed from the mind" would be developed by Neoplatonists such as Plotinus into a simile for intellective self-awareness.[25] Augustine explained that the mind is a mirror of God, one that it is necessary to both look upon and look through.[26] Effectively, the glass metaphor captures the limitations of human mentality: like the lowly mind, a mirror can only receive light, not create it; its images are fleeting; it has no understanding of the thing it reflects; and its pictures are mere shadows of the reality.[27]

Bolstered no doubt by the establishment of the glassworking industry from the twelfth century on, the Neoplatonic mirror grew into a worldview, an idea that God's image is reflected at all stages of the creational hierarchy, the cosmic "chain of mirrors."[28] Texts too were mirrors, as is apparent from the explosion of books bearing the word *speculum*—later, *mirror* and *glass*—in their titles. These mirror texts, notes Herbert Grabes, generally show things as they are, as they should be, or as they will be; there seems as well to be a fourth category, comprising texts that present what only exists "in the writer's imagination."[29] Seventeenth-century mirror texts do not merely reflect but purposefully and artfully distort: increasingly, they called themselves "optic," "perspective," or "prospective" glasses, as we see, for example, with Thomas Walkington's 1607 treatise on the passions, *The Opticke glasse of hvmors*.[30]

---

24. Marsilio Ficino, *Three Books on Life*, trans. Carol V. Kaske and John R. Clark (Binghamton, NY: Medieval & Renaissance Texts & Studies in conjunction with the Renaissance Society of America, 1989), 161.

25. Plato, *Timaeus*, 71b. See also Plotinus, *Ennead, Volume I: Porphyry on the Life of Plotinus. Ennead I*, trans. A. H. Armstrong, Loeb Classical Library 440 (Cambridge, MA: Harvard University Press, 1969), 1.4.10, which says there exists "a smooth, bright, untroubled surface" within us that "mirrors the images of thought and intellect."

26. This is Augustine's understanding of 1 Corinthians 13:12, "We now see through a mirror." See *On the Trinity*, ed. Gareth B. Matthews, trans. Stephen McKenna (Cambridge: Cambridge University Press, 2002), 214.

27. Herbert Grabes, *The Mutable Glass: Mirror-Imagery in Titles and Texts of the Middle Ages and English Renaissance* (Cambridge: Cambridge University Press, 1982), 85.

28. Benjamin Goldberg, *The Mirror and Man* (Charlottesville: University Press of Virginia, 1985), 116. On the history of glassmaking in Europe, see Sabine Melchior-Bonnet, *The Mirror: A History*, trans. Katharine H. Jewett (New York: Routledge, 2001), 9–100.

29. Grabes, *Mutable Glass*, 39, 62–63, 82–93.

30. Some examples: John Vicars, *A Prospectiue glasse to looke into Heauen* (1618); Edward Cooke, *The Prospectiue glasse of warre* (1628); John Philpot, *A Prospective-glasse for gamesters* (1646); Anthony Saint Leger, *A Prospective glass for the revolters at sea* (1648); Samuel Hinde, *Englands prospective-glass* (1663).

Keen interest in catoptric illusions is palpable among early modern natural philosophers. Earlier mathematicians had with carefully notated diagrams explained how, for example, a concave lens could distort the seeming magnitude and proximity of objects so that "what lies toward the top will be seen toward the bottom" and "right-hand things sometimes appear to the left."[31] The Renaissance attitude toward distortive glass is more wondrous. The occultist Heinrich Cornelius Agrippa speaks of lenses that can kindle fire from sunbeams and summon all the colors of the rainbow.[32] Cardano writes of glasses that can spy objects "five miles" away, even "with walls intervening," or illumine "dim and deep-down things, such as the womb, the throat, or a dark room."[33] Porta describes how "Catoptrick-glasses" may produce deformed faces or images floating in air. What could be more marvelous, he asks, "then that certain experiments should follow the imaginary conceits of the mind," fashioning illusions that seem not "the repercussion of the Glasses, but Spirits of vain Phantasms?"[34] Science, it seems, could now imitate and even outdo the holographic effects of the image-making faculty.

Indeed, as it struck many, imagination itself rather behaves like a cunning lens. In *The Art of English Poesy*, George Puttenham writes that the "fantasy" is like a "glass," of which "there be many tempers and manner of makings, as the perspectives do acknowledge, for some be false glasses and show things otherwise than they be indeed, and others right as they be indeed, neither fairer nor fouler, nor greater nor smaller. There be again of these glasses that show things exceeding fair and comely, others that show figures very monstrous and ill-favored."[35] As Rayna Kalas has argued, the arrival of clear *cristallo* glass in this period was concomitant with a new assessment of poesy as "an optic like the eye, rather than a series of pictures conjured up by words." Accordingly, Puttenham's analogy registers that imagination is "as artificial as the instrumental technology of a perspective glass."[36] However, poetry was not the only context in which imagination was instrumentalized: contemporary explana-

31. Ptolemy, *Ptolemy's Theory of Visual Perception: An English Translation of the "Optics,"* trans. A. Mark Smith (Independence Square, PA: American Philosophical Society, 1996), 216.

32. Henry Cornelius Agrippa, *Of the Vanitie and Vncertaintie of Artes and Sciences*, ed. Catherine M. Dunn (Northridge: California State University, 1974), 83–84.

33. Cardano, *"De Subtilitate,"* 246, 249–52.

34. Porta, *Natural magick*, 355.

35. George Puttenham, *The Art of English Poesy*, ed. Frank Whigham and Wayne A. Rebhorn (Ithaca, NY: Cornell University Press, 2007), 110.

36. Rayna Kalas, *Frame, Glass, Verse: The Technology of Poetic Invention in the English Renaissance* (Ithaca, NY: Cornell University Press, 2007), 2–3, 138. See also Kalas's distinction between the "decorously imperfect" reflections of the steel glass and the more patently "false" ones of the glassy mirror, which evokes the comparable contrast often drawn between the icastic and fantastic imagination (116).

tions of cognition often present the fantasy as an optical tool. Such accounts may be seen as attempts to take the measure of imagination's routine distortions, as geometers had done with the eye—to normalize the errors of the mind, as it were, to mathematize insight. The fifteenth-century reformer and preacher Girolamo Savonarola, for example, advised his listeners to employ imagination as "spectacles of death." Just as a corrective lens may be placed between the eye and the page in order that the reader may better make out the text, so "mental images" of mortality may usefully mediate between the transitory world of sensible things and the intellect.[37] Medieval and early modern Neoplatonism saw *phantasia* as dangerous, capable of corrupting the soul; at the same time, though, imagination promises self-knowledge.[38] As one sixteenth-century poet describes, "*Fancies* Glasse" makes our immortal nature visible before itself.[39] It is by "looking into *Fancies* Mirrour" that the soul judges her own health—a sick conscience, for example, fills the glass with monsters, with "grimme *Chimaeraes*" and "*sights* full of *horror.*"[40]

Moral and spiritual matters aside, optical tropes supply a way of understanding how the different faculties collaborate. Fulke Greville calls "imagination" an "organ" of "knowledge," designed especially to serve as the adjunct of the higher mind. It is

> a glasse, wherein the obiect of our Sense
> Ought to respect true height, or declination,
> For vnderstandings cleares intelligence:
> But this power also hath her variation,
> Fixed in some, in some with difference;
> In all, so shadowed with selfe-application
> As makes her pictures still too foule, or faire;
> Not like the life in lineament, or ayre.[41]

Imagination may be given to inconsistency ("fixed in some, in some with difference") and predisposed to recursive solipsism ("selfe-application"); yet it remains the necessary intermediary between "Sense" and "vnderstanding." These ideas are confirmed by the poet John Davies, who writes similarly that

---

37. Girolamo Savonarola, *Selected Writings of Girolamo Savonarola: Religion and Politics, 1490–1498,* trans. and ed. Anne Borelli and Maria Pastore Passaro (New Haven, CT: Yale University Press, 2006), 43–46.

38. Gerard Watson, *Phantasia in Classical Thought* (Galway: Galway University Press, 1988), 96.

39. John Davies of Hereford, *Microcosmos* (Oxford, 1603), 233.

40. John Davies of Hereford, *The Muses sacrifice* (London, 1612), 166v.

41. Fulke Greville, "A Treatie of Humane Learning," in *Certaine learned and elegant workes* (London, 1633), 24.

"Phantasie" is "wits looking glasse," a "mirrour" with which the "vnderstanding" may examine "the gatherings of the Senses."[42] The fantasy here is not
simply a symbol of the mind's meager power; it is part of a cognitive system
whose operations and errors may be understood as coherently as eclipses in
the sky. Accordingly, a mix of celestial and optical tropes appears in Philippe
de Mornay's defense of Christianity, a text that styles itself as a geometric proof
of the religion. It describes how the mind "beholdeth and seeth through his
imaginations as it were through a Clowde, [and] is after a sort trubbled by the
dimming of the Spectacles and by the smoakinesse of the imaginations."[43]
Like cloud and smoke, imagination disperses light. Like a lens, it bends light.
Where the passions are involved, fancy's glass acts as a filter: Thomas Wright
notes that oftentimes imagination "putteth greene spectacles before the eyes
of our witte, to make it see nothing but greene," coloring understanding
with emotion.[44] The foregoing examples do not simply equate fancy with
mental fallibility; they attempt with some specificity to take the measure of
imagination's normal distortions, to grasp both its functionality and its routine errors with something approaching mathematical certainty.

Thus, if at the turn of the seventeenth century the outward eye metamor-
phosed into a camera obscura, the inward eye also became, in its own way,
akin to an optical device—an idiosyncratic cognitive prosthesis, possessing its
own quirks. The similarity had survived, in a sense, though it was subtly transformed. Strictly speaking, the eye is not merely like a camera; it is exactly a
camera—just as the heart is a pump and a joint is a lever. There may have been
speculation that the "light, and cleernesse" of imagination had a physiological basis: melancholy, we will see in the next chapter, had the power to darken
the mind.[45] For the most part, though, the unverifiable optics of imagination
did not escape the realm of the figurative. Seeing, meanwhile, was newly privileged in the new empirical age. For early modern theologians, bodily sight is
the poorer cousin of insight. The reverse is true among physicians and engineers: when they call the eye a "mirror or Looking-glasse," they mean to stress

---

42. John Davies, *Nosce teipsum* (London, 1599), 46, 49.

43. Philippe de Mornay, seigneur du Plessis-Marly, *A woorke concerning the trewnesse of the Christian religion*, trans. Philip Sidney and Arthur Golding (London, 1587), ***6r, 254.

44. Thomas Wright, *The passions of the minde in generall* (London, 1604), 51.

45. Juan Huarte, in *The examination of mens wits*, trans. Richard Carew (London, 1594), 233, says
that just "as the eies stand in need of light, and cleernesse, to see figures and colours, so the imagination hath need of light in the brain, to see the fantasies"; this light emanates from "the vital spirits
which are bred in the heart." See also Timothy Bright, *A treatise of melancholie* (London, 1586), 103,
which describes how the melancholic humor creates "internall darkness" and "substantiall obscurity"
in the body.

the rectitude of its perception, the beauty of its material architecture.[46] André du Laurens calls the eye a "beautifull christall" possessing an "infallible sence" that is rivaled only by the faculty of understanding. For Cardano as well, vision is on par with the intellect.[47] Leonardo's praise is perhaps the most fulsome: for him, the eye is the author of astronomy, cosmography, mathematics, architecture, and painting; it is the great nullifier of "confusions and lies."[48]

That being said, Leonardo also confessed a habit of exercising his imagination at night, in darkness running his mind's eye over "the surface delineations of forms previously studied" so that he might better remember them.[49] Francis Bacon, who declared in his *Great Instauration*, "I admit nothing but on the faith of the eyes," also wrote that "God hath framed the minde of man as a mirror or glass capable of the image of the universal world."[50] The sixteenth-century English edition of Euclid's *Geometrie* repeatedly instructs its reader to mentally confirm the truth of its theorems with the visualizing aid of "imagination."[51] For Kepler, the abstraction of mathematics was itself a form of sight: without it, he wrote, "I am like a blind man."[52] The yoke between eyesight and insight had not broken: one is true, the other a necessary approximation of the truth. But it is not always immediately apparent which is which.

TRUEST SIGHT BEGUILED

A key result of Keplerian optics is that human vision is always inverted: rays of light crisscross in the eye lens, and we see the world upside down. This ocular chiasmus is analogous to the rhetorical device of paradox, which the early moderns associated with profound wisdom.[53] This wisdom is certainly the premise of Charles Estienne's *Defence of contraries*, which argues in favor of various contradictory propositions, including the following:

46. See, for example, Helkiah Crooke, *Mikrokosmographia* (London, 1615), 699.

47. André du Laurens, *A discourse of the preseruation of the sight*, trans. Richard Surphlet (London, 1599), 17, 32; Cardano, "De Subtilitate," 702. Du Laurens's treatise was published earlier as *Discours de la conservation de la veue* (Tours, 1594).

48. Leonardo da Vinci, *Leonardo on Painting*, trans. Martin Kemp and Margaret Walker, ed. Martin Kemp (New Haven, CT: Yale University Press, 1989), 21–22.

49. Leonardo, 224–25.

50. Francis Bacon, *The Works of Francis Bacon*, ed. James Spedding, Robert Leslie Ellis, and Douglas Denon Heath (London: Longman, 1857) 3:265, 4:30.

51. Euclid, *Geometrie*; see, for example, 318v, 326v, 341r, 343v, 363v.

52. *Sine quibus ego coecus sum.* Johannes Kepler, *Apologia pro opere harmonices mvndi*, in *Werke*, 6:397.

53. Galen gave the name *chiasma* to the diagonal crossing point of the two optic nerves, thought to resemble the Greek letter *chi*, written as X. See Siegel, *Galen on Sense Perception*, 59–60.

For Pouertie. That it is better to be poore than Rich.

For Blindnesse. That it is better to be blinde than to see cleerely.

For the Foole. That it is better to be a foole, than wise.

For him that hath lost his worldlie goods, honors and Preferments. That a man ought not to bee greeued, though he be dispoiled of all his goods and Honours.

For the Exiled. That it is better to be bannished, then continue in libertie.

For the Bastard. That the Bastard is more to be esteemed, than the lawfully borne or legitimate.[54]

Estienne's work, translated into English by Anthony Munday in 1593, derives from Ortensio Lando's *Paradossi* of 1543, which in turn was modeled on Cicero's *Paradoxa stoicorum.*[55] It belongs to what Rosalie Colie describes as a Renaissance vogue—"exercises of wit designed to amuse an audience sufficiently sophisticated in the arts of language to understand them."[56] Shakespeare's *King Lear* is one such exercise, clearly—for Estienne's list reads virtually as the play's dramatis personae, recalling, in order, Poor Tom, blind Gloucester, wise Fool, unaccommodated Lear, exiled Kent, and illegitimate Edmund. The principle of contrariety is acknowledged by Shakespeare's characters, who repeatedly observe that theirs is an age of reversal, a time when "old fools are babes again" and the father is "child-changed" (Goneril, 1.3.15.4; Cordelia, 4.7.16), when "the cart draws the horse" (Fool, 1.4.193), "smiling rogues" abound (Kent, 2.2.66), and "ruinous disorders" are the norm (Gloucester, 1.2.102).[57]

In the critical tradition of *King Lear*, it is the paradox of blind insight that has been perennially singled out. The play's fraught representation of seeing, epitomized in the gruesome blinding of Gloucester, has been read as a calculus of tragic compensation, wherein loss of vision results in a gain in wisdom. Those "who pride themselves on their clear-sighted appraisal of the world and its ways, find themselves betrayed by their sight," while "those who have lost their eyes may, in the very moment of losing them, receive a flash of moral

54. Charles Estienne, *The Defence of contraries*, trans. Anthony Munday (London, 1593), O2v–O3r. Other chapters deal with "the ignorant," "the infirmitie of the body," "teares," "dearth," and "hard lodging." The bastardy chapter was planned for the second volume.

55. Patrizia Grimaldi Pizzorno, *The Ways of Paradox from Lando to Donne* (Florence: Leo S. Olschki, 2007), 18.

56. Rosalie Colie, *Paradoxia Epidemica: The Renaissance Tradition of Paradox* (Princeton, NJ: Princeton University Press, 1966), 4.

57. For more on paradox, see Peter Stallybrass and Allon White, *The Politics and Poetics of Transgression* (London: Methuen, 1986); Patricia Parker, *Shakespeare from the Margins: Language, Culture, Context* (Chicago: University of Chicago Press, 1996), chap. 1, "Preposterous Estates, Preposterous Events"; and Peter G. Platt, *Shakespeare and the Culture of Paradox* (Farnham, England: Ashgate, 2009).

illumination."[58] Such a reading would have stood in Shakespeare's own time, as Estienne's essay on blindness suggests: it tells the story of how the philosopher Democritus supposedly destroyed his eyesight by staring at the sun, so that "he might recouer better vse of the eies of the minde." Estienne glorifies the blind man's "strength of spirit, better imagination, and contemplation of things high and heauenly." The blind have "no need of spectacles"; they remain untroubled by ocular ailments such as "the web, pearle, [and] teares fistula."[59] Comparable arguments about the happy lot of the blind were made by others around the same time.[60]

But maybe it is more accurate to say that *King Lear* moves beyond what was in 1606 an already formulaic trope, demonstrating instead that our notions of sanity, rationality, or goodness are tacitly in dialogue with our theories of sensation and perception. If it is believed that the eye is an insentient instrument whose distortions the mind is sometimes, but not always, able to correct, then neither eye nor mind cleanly attaches to the idea of acuity. If the presumed resemblance between eyesight and insight is eroded, it would mean that tropes derived from that resemblance—such as the inversion of that resemblance—will be depleted in force also. In its final stages, *King Lear* gathers its tragic potency not from the chiastic substitution of imagination for vision, or even from the epic miscarriage of that substitution, but rather from the elegiac dismantling of the presumed chiastic structure itself. To see this, we might take *Venus and Adonis* as a counterpoint or even a prelude to the tragedy. Written over a decade earlier, the epyllion portrays a rupture between mental and visual perception that leads to the making of a contrarious world. *King Lear* takes such a world as its premise. In the poem, Shakespeare's thinking captures the state of early modern optics and represents the separation of looking and imagining. But the play goes further, tearing at the figurative resemblance between the two even as Renaissance psychologists were trying to uphold it, and deepening Lear's downfall by tying his demise to the crumbling of a long-standing philosophical paradigm.

In his retelling of Ovid, Shakespeare departs from his source by spotlighting the relation between Venus's eyes and mind. In the *Metamorphoses*, Venus's

58. Derek Traversi, *An Approach to Shakespeare* (London: Sands, 1957), 203. For an expanded version of this argument, see Robert Bechtold Heilman, *This Great Stage: Image and Structure in "King Lear"* (Baton Rouge: University of Louisiana Press, 1948), 41–66.

59. Estienne, *Defence of contraries*, 33, 34, 38–39.

60. See, for example, Pierre de la Primaudaye, *The second part of the French academie*, trans. Thomas Bowes (London, 1594), 79, which says the impious man might as well pluck out his eyes, so that he is "blinde both in body and soule"; and George Hakewill, *The vanitie of the eye* (Oxford, 1608), 152, which calls blind men the moral "lamps and torches in the world."

reaction to the death of Adonis is passionate but relatively brief: "shee from the skye / Beehill[s] him dead," and proceeds to

> tare at once hir garments from her brist,
> And rent her heare, and beate vppon her stomack with her fist.[61]

Shakespeare, in contrast, gives us an extended sequence of perceptual transformation, wherein the goddess is aggrieved first in her mind and then in her eyes and is further dismayed by the conflict between the two. Initially, Venus does not see Adonis, only hears the lamenting howls of his hunting dogs. This in itself is enough to immobilize her with "causeless fantasy" (897). Her stunned "drunken brain" begins to swim with apprehensions, and Venus weeps:

> Her eyes seen in the tears, tears in her eye:
> Both crystals, where they viewed each other's sorrow—(962–63)

As in sonnet 43, chiasmus is used to suggest an optical effect—here, the watery embrace of eyes and tears, each said to "lend and borrow" light. Also on display is the usual imagery of precious materials: Venus weeps a crystal "tide"; her tears are "prisoned in her eye, like pearls in glass" (979–80). Soon, Adonis's mauled body comes into sight, and the goddess's eyes are "murdered with the view," swiftly withdrawing their light, "like stars ashamed of day" (1031–32).

Shakespeare might have ended here, with an image that, grandly but conventionally, links Venus's gaze to the light of celestial spheres. But the simile continues into the next stanza:

> Or as the snail, whose tender horns being hit,
> Shrinks backwards in his shelly cave with pain,
> And there, all smothered up, in shade doth sit,
> Long after fearing to creep forth again:
>   So at his bloody view her eyes are fled
>   Into the deep-dark cabins of her head. (1033–38)

From winking eyes to creeping snail: bizarrely, the poet translates the eyes of his divine heroine into the viscid horns of a gastropod. The metaphor is as unexpected as it is grotesque, for snails are more often symbols of slowness—

---

61. Ovid, *The xv. bookes of P. Ouidius Naso, entytuled Metamorphosis,* trans. Arthur Golding (London, 1567), 134r.

recall Jacques's schoolboy, "creeping like snail / Unwillingly to school" (*As You Like It*, 2.7.146–47). Interminably, the trope inches, over the course of four lines, from "being hit" to "shrinks backwards," to "smothered up," "doth sit," and "creep forth again." Possibly, the point here is mimetic: the metaphor makes it hard for us to envisage Venus, as hard as it is for her to look on dead Adonis.[62] And yet, the image's outlandishness defeats empathy.[63] The more immediate effect is to call attention to our entering into the body, that murky territory where Kepler would later decline to tread. Like the imploring Eyes of the *Paralipomena*'s epigram—*deliver us from darkness*—Venus's eyes

> resign their office and their light
> To the disposing of her troubled brain. (1039–40)

But here is no hospitable meeting between sense and intellect. Eyes are angrily repulsed, driven out "from their dark beds once more" (1050). Venus again "thr[ows] unwilling light" on the corpse (1051):

> Upon his hurt she looks so steadfastly
> That her sight, dazzling, makes the wound seem three;
> And then she reprehends her mangling eye
> That makes more gashes where no breach should be:
> His face seems twain; each several limb is doubled;
> For oft the eye mistakes, the brain being troubled. (1063–68)

Seeing doubly and triply, Venus's "mangling eye" "makes more gashes" on the body of her paramour. *Dazzle* carries a dual sense of injury and injuring—for eyes may "dazzle" both when they are blinded and when they blind.[64] Elsewhere in Shakespeare, tears are able to fracture vision:

> Sorrow's eye, glazed with blinding tears,
> Divides one thing entire to many objects,
> Like perspectives. (*Richard II*, 2.2.16–18)

But Venus is now dry-eyed, finding her "salt tears gone" (1071); it is her troubled "brain" that is to blame for her newly splintered view.

---

62. Richard Meek, *Narrating the Visual in Shakespeare* (Farnham, England: Ashgate, 2009), 47.
63. Eugene B. Cantelupe, "An Iconographical Interpretation of *Venus and Adonis*, Shakespeare's Ovidian Comedy," *Shakespeare Quarterly* 14, no. 2 (1963): 147.
64. *OED Online*, s.v. "dazzle, v." 1 and 3 (Oxford: Oxford University Press, January 2018). http://www.oed.com/view/Entry/47588.

Marcus Nordlund has argued that *Venus and Adonis* exemplifies what he calls the "transitional" visual politics of the Renaissance, the tension between emissive and intromissive theories of seeing. The eager goddess models sight as "an active and participational extension of the self," while the sullen boy suggests "strictly passive and neutral reception."[65] These two models, Eric Langley adds, suggest different forms of subjectivity, mutuality versus narcissism.[66] Such readings certainly illuminate the moment when Venus lifts Adonis's eyelids, finding "two lamps burnt out in darkness"; in the next instant, lamps turn into mirrors, "two glasses where herself herself beheld" (1128–29). But to focus solely on the ocular story means to overlook the imaginative one. Shakespeare endows his Venus with fantastical vigor. Earlier in the poem, her inner "eye" is visited by a premonitory "picture" of the "chafing boar" and an "image" of Adonis "stained with gore"—a terrifying "imagination" (661–68). Earlier still, her erotic frustration is like that of "poor birds, deceived with painted grapes"; the seeming signs of Adonis's ardor, she learns, are but "imaginary" Zeuxian illusions (597, 601). When Adonis takes his leave, the light in Venus is "blown out"; "confounded in the dark," she descends into a "fantastic" "humor," wailing a "woeful ditty" (826–27, 836, 850). The power of Venus's imagination is such that it can infuse paintings with life, see forward in time, adorn her eyes with teary jewels. But Adonis's death opens up a permanent incongruity between her mind and her sense, one that she takes to be part of a larger schism in nature. "Wonder of time," she grieves, looking at his now darkened eyes, that "thou being dead, the day should yet be light" (1133–34).

Venus then moves to codify her perceptual reorientation in the form of an oath. In Ovid, the goddess drips divine nectar into Adonis's blood, thus creating the purple flower that will become his living commemoration. In Shakespeare, the flower seems almost incidental: it springs up, and Venus plucks it. The real metamorphosis has come before this—not of Adonis but of love itself, which Venus determines will forevermore be a thing of contradictions and paradoxes, will "the truest sight beguile" (1144). Love will be "raging mad and silly mild," "merciful and too severe." It will "strike the wise dumb, and teach the fool to speak"; "make the young old, the old become a child"; be "cause of war and dire events, / And set dissension twixt the son and sire"

---

65. Marcus Nordlund, *The Dark Lantern: A Historical Study of Sight in Shakespeare, Webster, and Middleton* (Göteborg, Sweden: Acta Universitatis Gothoburgensis, 1999), viii–ix.

66. Eric Langley, "'And Died to Kiss His Shadow': The Narcissistic Gaze in Shakespeare's *Venus and Adonis*," *Modern Language Studies* 44, no. 1 (2008): 12–26. See also John McGee, "Shakespeare's Narcissus: Omnipresent Love in *Venus and Adonis*," *Shakespeare Survey* 63 (2003): 272–81, which argues that Venus outdoes Adonis in self-adoration.

(1146–60). This is the world of *King Lear*, of course. Whereas in *Venus and Adonis* a visual trauma births paradox, Shakespeare's tragedy at the outset assumes that things are not as they seem, cannot seem as they are.

SEEMING TO SEE

Rather than heal the rift between mind and sense, *King Lear* widens it. Relegating corporeal sight to banality on the one hand, the play interrogates imagination's ability to compensate for failures in vision on the other. Its most cynical comment on the eye is voiced in the instant that Cornwall wrests Gloucester's second eye from its socket:

> Out, vile jelly!
> Where is thy luster now? (3.7.83–84)

The word *vile* appears several times in *King Lear*, usually in paradoxical contexts: for example, Lear says that "the art of our necessities is strange," for it "can make vile things precious" (3.2.70–71). "Jelly" is harder to explain. Hitherto used for foodstuffs, the word had only recently come to mean a type of consistency.[67] Jelly is not among the analogical constituents associated with the anatomical eye: the albuminous corneal humor, "like vnto the white of an egge"; the crystalline lens, like a "steele-glasse" or "spectacles"; the vitreous humor inside the eyeball, akin to "moulten glasse."[68] On the rare occasions that *jelly* comes up in connection with eyes, it is to indicate illness: Estienne lists "eye-gellie" as an ocular malady, for instance.[69] It seems, therefore, that Cornwall's "vile jelly," rather like Venus's molluscan stalks, is meant to shock—by reducing the eye's lucent architecture to formless slime. Just a few lines later, there is a different irony: a servant fetches "some flax and whites of eggs to apply to [Gloucester's] bleeding face," (3.7.99.8–9), invoking the common use of egg whites as a topical treatment for eye problems.[70] For this patient, whose

---

67. *OED Online*, s.v. "jelly, n.1" 2 (Oxford: Oxford University Press, January 2018). http://www.oed.com/view/Entry/101023.

68. Du Laurens, *Discourse*, 33–34.

69. Estienne, *Defence of contraries*, 39. See also Guillaume de Salluste du Bartas, *Du Bartas his deuine weekes and workes translated*, trans. Josuah Sylvester (London, 1611), Ii4v, which calls "Catharact" a "tough humour like gelly."

70. These lines appear only in the Quarto text, 3.7.99.8–9. For medical corroborations, see, for example, Thomas Cartwright, *An hospitall for the diseased* (London, 1580), 38; Philip Barrough, *The methode of phisicke* (London, 1583), 40; and Walter Bailey, *Two treatises concerning the preseruation of eie-sight* (Oxford, 1616), 53.

eye sockets no longer contain any albuminous humor, this medically accurate salve seems especially paltry.

Such clinical precision is of a piece with the play's general representation of the human eye as an ineffectual, even expendable organ. Vision is poor in *King Lear*: Goneril points out that Regan, in her adulterous lust for Edmund, is looking "asquint" (5.3.66). Gloucester must fumble for his glasses before he can read Edmund's forged letter: "If it be nothing, I shall not need spectacles" (1.2.34–35). Anguish is an "eye" to be closed up; the wind an "eyeless rage" that destroys the things it catches (4.4.14, 3.1.7.1.2). For a daughter to say she loves her father "dearer than eyesight" means little (1.1.54). If Cornwall decides to "set my foot" on Gloucester's eyes (3.7.68), it is because Gloucester first imagined Regan plucking out Lear's (3.7.56–57). Lear himself is only too willing to part with his "old fond eyes":

Beweep this cause again, I'll pluck ye out
And cast you with the waters that you lose
To temper clay. (1.4.272–74)

Lear returns to this strange impulse again later, thinking he might "use his eyes for garden water-pots" (4.6.190). In this play, it seems eyes are not the instruments of the soul; rather, they are base materials in need of a productive purpose.[71]

It comes as no surprise, then, that the gouging of Gloucester's eyes is styled less as a cathartic awakening than as a bloody, almost farcical defilement. As Paul Alpers points out, the torture scene is unredeemed by visionary understanding.[72] The blinded man's first words verge on a kind of momentousness—"All dark and comfortless?" (3.7.85)—but, with his next breath, he prays that Edmund, the son who betrayed him, may avenge him. As he makes his way to Dover, Gloucester utters enigmatic paradoxes:

I have no way and therefore want no eyes.
I stumbled when I saw. (4.1.19–20)

Yet, at the same time, he clings to his outward senses, even as those very senses are undermined by Edgar, who convinces his father that he is unable to tell

---

71. Compare with La Primaudaye, *Second part of the French academie*, 79: "What is an eye pluckt out of the head but a litle clay and mire?"

72. Paul Alpers, "*King Lear* and the Theory of the 'Sight Pattern,'" in *In Defense of Reading: A Reader's Approach to Literary Criticism*, ed. Reuben A. Brower and Richard Poirier (New York: E. P. Button, 1962), 137.

"even" ground from "steep," or hear the nearby sea. If Gloucester becomes a sort of visionary, it is only because Edgar rouses his imagination with this extraordinary description:

> Here's the place. Stand still. How fearful
> And dizzy 'tis to cast one's eyes so low!
> The crows and choughs that wing the midway air
> Show scarce so gross as beetles. Halfway down
> Hangs one that gathers samphire: dreadful trade!
> Methinks he seems no bigger than his head.
> The fishermen that walked upon the beach
> Appear like mice, and yond tall anchoring bark
> Diminished to her cock; her cock a buoy
> Almost too small for sight. The murmuring surge,
> That on th'unnumbered idle pebbles chafes,
> Cannot be heard so high. (4.6.12–23)

In the mathematical precision of this picture, readers have found the signs of perspectivist art. Jonathan Goldberg notices, for example, "a kind of algebra that expresses a verbal version of a formula of proportion, a:b::b:c," with "yon anchoring bark" reduced to "her cock," and the cock further shrunken to "a buoy / Almost too small for sight."[73] Through these diminishing ratios, the eye "proportionately reduces and finally loses its subject in a distance."[74] If we look closely, though, the movement of the speech does not exactly give the sense of spatial regression. It roams, rather: we start in the "midway air," then drop "halfway down" the incline. We alight on the fishermen on the "beach," then drift outward to the "anchoring bark" at sea.

That is of course because the scene Edgar describes is not there at all. His speech is not a rendition of perspectival art but a sly imitation of it. If perspective portrays a three-dimensional reality by skewing lines on a two-dimensional surface, here imagination skews perspective, for Edgar discerns details that would be difficult, if not impossible, to make out in actuality. Minute "choughs" are recognized at great distance; it is known that the far-off climber gathers "samphire." The men on the beach are understood to be fishermen; the tall ship is invisibly anchored. To give the sense, felt by Simon Palfrey, that the "world is large and alive with small things," filled with "the possibility of

73. Jonathan Goldberg, "Dover Cliff and the Conditions of Representation: *King Lear* 4:6 in Perspective," *Poetics Today* 5, no. 3 (1984): 541–42.

74. Maurice Hunt, "Perspectivism in *King Lear* and *Cymbeline*," *Studies in the Humanities* 14, no. 1 (1987): 21.

magnification,"[75] Edgar varies the scale. The logic tying these images is aural
and associative as much as visual: after real birds come figurative "beetles."
Then we are told to imagine "one" that "hangs"—but the "one" in question is
a man, not an insect: the samphire gatherer. This man then telescopically disappears into his "own head." Similarly, fishermen turn into "mice," but
"cock," which comes after, refers not to the bird but to a boat. These rest-
less mutations, whereby we are shunted between animals, people, and objects, evoke the motions of the imagining mind, summoning half-formed
shadow meanings that rise up only to melt away momentarily. Over the visual errors and shrinking magnitudes that would deceive a physical eye, Edgar layers distortions that appeal particularly to the mind's eye.

After the fall, Edgar supplies still more mathematical information and more
details chosen to invoke imagination surreptitiously. He gives the "altitude"
and angle of Gloucester's leap—"ten masts"; "perpendicularly" (4.6.55–56).
The height of the "chalky bourn" is such that "the shrill-gorged lark so far /
Cannot be seen or heard" (4.6.60–61)—again, Edgar summons a bird only to
erase its image and muffle its cry. To one who had truly fallen such a height, such
details would presumably be of little interest; they can only appeal to one who
imagines them. Edgar's delusive ruse can be seen as both a compassionate ther-
apy, and "a theater of cruelty."[76] Either way, it functions with the aid of a phantasmatic vista, a scene that blends the ordinary human perspective with that of
an all-seeing eye, annotating verisimilitude with imagination. The effect is a kind
of complete vision that approximates, perhaps deliberately, that of the heavenly
geometer, expanding "human consciousness into a vessel for the divine."[77] As
though responding to the metaphysical cues, Gloucester dedicates the spectacle of his leap to the "sights" of the "gods" (4.6.36–37). The gods, of course,
are we. This salvific unification of vision and imagination has been wrought
by the playwright and enacted in the minds of his audience. If a revelation
has occurred, it is bestowed not by the blind man on us but rather by us on him.

Though the scene might have stopped at this uplifting juncture, it does not.
Instead, one set of cognitive illusions is followed immediately by another. With
the announcement "I am the King himself" enters Lear, addressing cryptic non
sequiturs to imaginary interlocutors (4.6.85–86). His mind, we discover, is already crammed with images:

75. Simon Palfrey, *Poor Tom: Living "King Lear"* (Chicago: University of Chicago Press, 2014),
173–74.

76. See, respectively, Stephen X. Mead, "Shakespeare's Play with Perspective: Sonnet 24, *Hamlet*,
Lear," *Studies in Philology* 109, no. 3 (2012): 254, and Andrew Bozio, "Embodied Thought and the Perception of Place in *King Lear*," *SEL Studies in English Literature 1500–1900* 55, no. 2 (2015): 271.

77. Erwin Panofsky, *Perspective as Symbolic Form* (Cambridge, MA: MIT Press, 1991), 72.

When I do stare, see how the subject quakes.
I pardon that man's life.—What was thy cause?
Adultery? Thou shalt not die. Die for adultery?
No, the wren goes to't, and the small gilded fly
Does lecher in my sight. Let copulation thrive,
For Gloucester's bastard son was kinder to his father
Than my daughters got 'tween the lawful sheets.
To't, luxury, pell-mell, for I lack soldiers.
Behold yond simp'ring dame,
Whose face between her forks presages snow,
That minces virtue and does shake the head
To hear of pleasure's name.
The fitchew nor the soiled horse goes to't
With a more riotous appetite. (4.6.108–21)

In a vague echo of Edgar's seascape, this phantasmatic harangue too mingles
the human and nonhuman: in among the adulterous men and the "simp'ring
dame" are the lecherous "wren" and "soiled horse." Instead of a perspectival
window, though, this is a moralizing mirror, trained on Boschian contortions
of earthly delight. The magnification is elastic and the view penetrative—Lear
sees copulating flies as clearly as the monstrous nether regions of women.
Whatever this picture represents—Lear's depraved kingdom, perhaps the
whole sorry lot of creation—its most important characteristic is total visibility, the fact that it is nakedly bare before the monarch's absolute "sight." Lear
has a comparable fancy in the earlier storm sequence, in fact, when he angrily
urges the "great gods" to root out their enemies:

> Tremble, thou wretch,
> That hast within thee undivulgèd crimes
> Unwhipped of justice. Hide thee, thou bloody hand,
> Thou perjured and thou simular man of virtue
> That art incestuous. Caitiff, to pieces shake,
> That under covert and convenient seeming
> Has practiced on man's life. (3.2.51–57)

Wrongdoing is characterized as "undivulgèd," "simular," and "covert." More
than the particular crimes in question—perjury, incest—Lear is incensed by
their concealment, by the audacity of the perpetrator who believes his guilt
can remain hidden from the sovereign. In the Quarto text, apparently inspired
to make his own enemies "tremble" and "hide," Lear stages a hallucinatory

arraignment of his daughters in the wind-beaten farmhouse. Determined to "see their trial," he summons Regan and Goneril to a phantom court before spectral men of law (3.6.15.19).

At the start of the play, what Lear sees and what he commands are identical: in the first scene he arrives with map in hand, as though literally to see his kingdom as he divides it among his offspring. When angered, he reflexively ejects Kent "out of my sight" (1.1.155), a move that entails banishment, a kind of nonexistence. As Kent tries in vain to point out, however, Lear's all-seeing gaze depends on the visual aid of his subjects. "See better, Lear," Kent says, "and let me still remain / The true blank of thine eye" (1.1.156–57). He is willing to be the "blank" or archer's target of Lear's wrath, maybe also the "blank" or white of Lear's own eye; either way, he offers to make himself the King's visual prosthesis.[78] Lear realizes his mistake only after he has alienated his friends and allies, and he is subsequently betrayed by Goneril:

> Does any here know me? This is not Lear.
> Does Lear walk thus? Speak thus? Where are his eyes? (1.4.195–96)

The lines may be played either as rhetorical hectoring or as genuine befuddlement, an initial step in Lear's sliding sanity. The ominous question, "Where are his eyes?" reveals how profoundly Lear invests his authority in an ideology of complete vision. Unable to see himself mirrored in his subjects, finding those around him unnervingly opaque, Lear deduces, not illogically, that he is "not Lear," that he has lost his "eyes." "Who is it that can tell me who I am?" he demands. "Lear's shadow," replies the Fool (1.4.199–200). A king without a kingdom is a shade without a body.

As his mind starts to unravel, Lear takes up the looking glass of imagination.[79] In its *speculum* he sees not his own soul but shadowy agents—a lascivious "beadle" lashing a "whore"; a duplicitous "usurer" who hangs his "cozener"; wearers of "robes and furred gowns" who "plate sins with gold" (4.6.154–59). These vile images support Debora Shuger's observation that early modern figurative mirrors were mostly showcases for good and bad exemplars, for generic assumptions about human nature and theological com-

---

78. Harry Berger, *Situated Utterances: Texts, Bodies, and Cultural Representations* (New York: Fordham University Press, 2005), 99–100.

79. If Allan R. Shickman's contention that the Fool carried a mirror on stage is to be believed, then the speculative nature of Lear's unraveling may have occurred to audiences all the more. See "The Fool's Mirror in *King Lear*," *English Literary Renaissance* 21, no. 1 (1991): 76–77.

SEEMING TO SEE 133

monplaces: "The object viewed in the mirror is almost never the self."[80] The
problem is that Lear's juridical omniscience is ethically defunct, accompanied by no desire to see justice served. "None does offend," he avers; "none, I
say, none" (4.6.162). To sit in judgment is impossible, perhaps, when vice is
universal. Or maybe it is that Lear's compassion cannot outweigh his revul-
sion. Too absorbed in his fancy's self-serving glass, Lear becomes a figure of
pathos: "Give me an ounce of civet," he pleads with an imaginary apothecary,
to "sweeten my imagination" (4.6.128–29).

The stage is now set for the meeting of blind Gloucester and mad Lear.
Harley Granville-Barker is among the earliest critics, certainly not the last, to
see this meeting of "the sensual man robbed of his eyes with the wilful man,
the light of his mind put out," as crucial.[81] This conference, with its implicit
symbolic symmetry—the mirroring of two types of sightlessness—promises
to embody on stage Estienne's paradox that "it is better to be blinde than to
see cleerely." We are expecting to see what Colie calls a dramatic "unmeta-
phoring," wherein we witness a literary cliché translated into actuality.[82] Given
what we have seen of the two men's compromised perception, however, the
emblematic potential is undercut from the start. In the taxing and sometimes
absurd conversation that ensues between these two blind seers, it is not their
similarity but their disjunction that emerges. Turning the screw, Shakespeare
steers the conversation explicitly toward the sensitive matter of looking.

GLOUCESTER. Dost thou know me?
LEAR. I remember thine eyes well enough. Dost thou squiny at me? No,
  do thy worst, blind Cupid. I'll not love. Read thou this challenge; mark
  but the penning of it.
GLOUCESTER. Were all the letters suns, I could not see.
. . . .
LEAR. Read.
GLOUCESTER. What, with the case of eyes?
LEAR. . . . Your eyes are in a heavy case, your purse in a light, yet you see
  how this world goes.
GLOUCESTER. I see it feelingly.
LEAR. What, art mad? A man may see how this world goes with no eyes.
  Look with thine ears. (4.6.133–37, 140–47)

---

80. Debora Shuger, "The 'I' of the Beholder: Renaissance Mirrors and the Reflexive Mind," in
Renaissance Culture and the Everyday, ed. Patricia Fumerton and Simon Hunt (Philadelphia: University
of Pennsylvania Press, 1999), 22.
81. Harley Granville-Barker, *Prefaces to Shakespeare* (London: Sidgwick & Jackson, 1927), 1:179.
82. Rosalie Colie, *Shakespeare's Living Art* (Princeton, NJ: Princeton University Press, 1974), 11.

The conversation is uncomfortable, and not only because of the dissonance between Lear's tactless levity ("Do thy worst, blind Cupid") and Gloucester's dolefulness ("I see it feelingly"). More subtly, the two speakers allegorize the insuperable rift between sense and imagination. The eyeless man is preoccupied only with his corporeal loss—"what, with the case of eyes?"—while the witless man raves about the mind's power—"yet you see how this world goes."

As the interview draws to its close, Lear takes a suggestive parting shot that seems to encapsulate this play's vexed view of seeing overall:

> Get thee glass eyes,
> And, like a scurvy politician,
> Seem to see the things thou dost not. (4.6.164–66)

In the simplest sense, Lear is saying that visual difficulties are easily resolved with "glass eyes"—Gloucester needs his spectacles. Read another way, the lines indict eyesight: eyes are only glassy ornaments, falsely advertising an acuity they do not possess. To have them is, at best, to "seem / To see." Or else, the point may be that imagination is to blame: like the hypocritical politician, the mind conceals itself behind eyes, seems "to see" while covertly indulging inward fabrications behind the "glass." In the premodern history of optics, vision was a graduated process of cognitive rarefaction, a transformation of things into mental figments. Lear's "glass eyes" instead imply a paradigm of perception that is premised on a ceaseless sequence of distortion and partial rectification. The choice now is not so much between seeing and not seeing as between seeming to see and seeming to see less wrongly.

Only as they part do the two men briefly make empathetic overtures. In a rare moment of lucidity, Lear offers to make reparations for Gloucester's loss:

> If thou wilt weep my fortunes, take my eyes.
> I know thee well enough; thy name is Gloucester. (4.6.170–71)

And Gloucester, though he remains sorrowfully sane—"how stiff is my vile sense" (4.6.269)—glimpses something of Lear's condition:

> Better I were distract;
> So should my thoughts be severed from my griefs,
> And woes by wrong imaginations lose
> The knowledge of themselves. (4.6.271–74)

SEEMING TO SEE   135

These are only gestures of conciliation. Lear offers to give away his eyes, which we know he would happily be rid of; Gloucester still thinks imaginations are "wrong." Ultimately, the scene recedes on a note of dissatisfaction. We have seen the blind man, heard the madman, but we have not witnessed the expected sublime synthesis of moral and intellective perspicacity. If anything, we sense that analogical structures are fated to eventually undo themselves. Even paradox, the enemy of *doxa*, is gradually immobilized, made moribund, through repetition.[83]

New reversals occur in the play's final scene. Unexpectedly, Cordelia dies. A mirror is put to unusual use: Lear, struggling to find proof of his daughter's life, tries to coax her breath onto a "looking-glass" (5.3.235). This reflective "stone" is meant not to show but rather to feel, to "mist" and "stain" with Cordelia's impossible exhalation; it is one more instance of *King Lear*'s ironic misuse of optical instruments. In its last moments, the play draws attention to itself as a theatrical representation. The tableau of Lear kneeling over Cordelia prompts Kent to ask helplessly, "Is this the promised end?" To which Edgar supplies a correction: "Or image of that horror" (5.3.237–38). But just as we are contemplating what distinguishes apocalyptic doom from its "image," the laws of visibility seem suddenly to change. Albany has begun what sounds like a concluding speech, apportioning just deserts to friends and foes, when he is surprised: "Oh, see, see!" (5.3.280). Lear, with his dying breath, directs us to some indiscernible event:

Do you see this? Look on her! Look, her lips,
Look there. Look there! (5.3.286–87)

Whatever their pleas are pointing to, it is surely significant that the play has excluded us from its wondrous conclusion. With our eyes trained on the actors, their reality is spirited away from us by Shakespeare's sleight of hand. We are left wondering not just what we have seen but also how we saw it.

King Lear thus deposits us on the threshold between literal vision and imaginative vision, the invisible frontier where one type of seeing supposedly accedes to another. It is one threshold among many that was negotiated by the early moderns as they sought to reconcile ancient and medieval philosophy, mathematics and technology, analogy and empiricism. Appropriating but also complicating broader themes embedded in the study of optics—likeness, error, instrumentation, representation—Shakespeare fashions a tragedy that seems to be less about enlightenment than about epistemological obscurity.

---

83. Platt, *Culture of Paradox*, 54.

The play shows that in the translation of flawed perception into perfect understanding, the image-making faculty bears an extraordinary, even impossible burden. As Renaissance mathematicians knew, a beam of light diverted from its course can only be corrected by diverting it anew. Not dissimilarly, the mind's eye emerges as a formidable yet flawed corrective for the newly mathematized corporeal eye; its distortive clarifications are only partly redemptive. If *King Lear* ends by mourning that we will never again see so much, it also asks what measure of blindness we accept in order to see clearly.

CHAPTER 5

Melancholy, Ecstasy, Phantasma
The Pathologies of *Macbeth*

A man with sharp, staring eyes sits in a room overrun with humming insects. Another, at the sickbed of an ailing friend, notices the vigor draining from his own body. A crazed courtier is visited by nightly visions of greed and debauchery. A young woman unhinged by a lover's rejection is nursed by indulging her delusions. These scenes, respectively from Edmund Spenser's *Faerie Queene*, Michel de Montaigne's essay on the power of imagination, John Marston's *Malcontent*, and Shakespeare's *Two Noble Kinsmen*, represent the kinds of stories that were told about the harmful power of the fantasy during the Renaissance. Together, physicians, natural philosophers, divines, and demonologists wrestled with the pathologies of the image-making faculty, a challenge that lay on several fronts. As melancholic complaints were becoming apparently widespread, so too was imaginative dysfunction, which emerged as melancholy's primary symptom. In unprecedented detail, early modern medical discourse attempted to map the relationship between problems in perception and the black bile. At the same time, cognitive phenomena such as dreams and hallucinations were taxonomized with increasing nuance as psychologists sought to separate carefully the realms of faculty psychology, humoral medicine, diabolical possession, and sin. This confluence of factors informs the representation of mental illness in Shakespeare's *Macbeth*, a play that does not merely reflect but rather interrogates the assumptions underpinning the medical thinking of its day. Three particular maladies featured in the tragedy—humoral melancholia, self-alienating ecstasy, and hallucinatory phantasms of criminal intent—challenge the presumed border between the interior and exterior, the native and the alien, sanity and disorder. More grimly than the comparably melancholy *Hamlet*, *Macbeth* broaches the unspoken anxiety undergirding contemporary clinical accounts: To what extent do our mental images belong to us? Shakespeare blurs the distinction between sick fantasies and ordinary subjectivity, uncovers the impossibility of diagnosing mental diseases as such, calls into question the very presumption of a mental interior, and so comments on the acute medical crisis that had seemingly come about at the turn of the seventeenth century.

Fantastical Melancholy

The fantasy could impair physical and mental health in myriad ways. We have seen in earlier chapters that imagination could muddle and become muddled by the physiological body; it was believed to be vulnerable to atmospheric and environmental influences; it was, we will see in the next chapter, nothing short of a disruptive force in nature itself. Yet it is in the context of perceptual disability and maladies of delusion that the faculty is perhaps most often discussed in early modern discourse. The case history of hallucination is remarkable for its size and age: stories testifying to the fantasy's ability to deceive body and mind alike crowd the pages of medical and spiritual texts. Montaigne's essay *De la force de l'imagination*, for example, gathers in a typical fashion a great many legends and folktales involving false perception, self-fulfilling superstitions, and crafty placebos. An about-to-be pardoned man falls dead at the mere sight of the scaffold; a bullfight spectator's delight causes him to grow horns overnight; a falconer summons his bird using only his eyes.[1] Pierre de la Primaudaye writes similarly of fantastical men who believe "that they haue a serpent, or some other beast in their bodies: others that they are become water pottes or glasses, and thereupon are afraide lest some body should iustle against them, and break them in peeces."[2] The theologian Ludwig Lavater tells of men who imagined that they were horned like oxen, or that they were dead, or else great princes and apostles.[3]

---

1. Michel de Montaigne, "Of the force of imagination," in *The essayes*, trans. John Florio (London, 1603), 40, 45.

2. Pierre de la Primaudaye, *The second part of the French academie*, trans. Thomas Bowes (London, 1594), 165.

3. Ludwig Lavater, *Of ghostes and spirites walking by nyght*, trans. Robert Harrison (London, 1572), 11. Lavater's *De spectris, lemuribus et magnis atque insolitis fragoribus* appeared first in 1569.

The kinds of *fabulae* that appear in the early modern literature of imagination arise as well in writings about melancholic disorder. The physicians Levinus Lemnius and Timothy Bright mention in their accounts of melancholy sufferers who believed themselves to have grown snouts or be missing their heads.[4] An extensive list of examples would be gathered in Robert Burton's *Anatomy of Melancholy*, in which we read of the man who "thinkes himselfe so little, that he can creepe into a mousehole," the one who "feares heaven will fall on his head," and one who "thinkes he is a Nightingall, and therefore sings all the night long."[5] As Michael MacDonald intriguingly notes, "None of the famous examples of melancholy delusion was native, and English doctors seem to have encountered very few fabulous fancies in practice."[6] The outlandish and archaic apocrypha of imaginative lore, in other words, constitute a discourse that was constructed as such. These stories also suggest the special way in which faculty psychology and humoral medicine amplified one another: the psychosomatic power of imagination, an idea already centuries old before the Renaissance, was underlined by the increasing prominence of melancholy in the medical commentary of the period. The fantasy's pathologies would become so vexed and complex, we will see, that they became harder, ironically, to recognize.

The coincidence of melancholy disease and perceptual disorder had been noted by the ancients. Aristotle found that confused and monstrous visions tend to appear in the dreams of the melancholic, feverish, and intoxicated.[7] Galen wrote that the psychosomatic force of the fantasy was most developed in "the case of people suffering from melancholy, phrenitis, or mania."[8] He surmised that melancholic imbalances lead to the involuntary creation of phantasms in the brain, which in turn lead to misperceptions and wrongful beliefs.[9] Similarly, Avicenna notes that "fear of things which do or do not exist" is among the major symptoms of melancholy; patients "imagine themselves

4. Levinus Lemnius, *The touchstone of complexions*, trans. Thomas Newton (London, 1576), 150v; Timothy Bright, *A treatise of melancholie* (London, 1586), 189.

5. Robert Burton, *The Anatomy of Melancholy*, ed. Thomas C. Faulkner, Nicolas K. Kiessling, and Rhonda L. Blair (Oxford: Clarendon, 1989), 1:402.

6. Michael MacDonald, *Mystical Bedlam: Madness, Anxiety, and Healing in Seventeenth-Century England* (Cambridge: Cambridge University Press, 1981), 154.

7. Aristotle, *On Dreams*, in *On the Soul. Parva Naturalia. On Breath*, trans. W. S. Hett, Loeb Classical Library 288 (Cambridge, MA: Harvard University Press, 1957), 461a14–25.

8. Galen, *Galen: Selected Works*, trans. P. N. Singer (Oxford: Oxford University Press, 1997), 160.

9. Galen, *On the Affected Parts*, trans. and ed. Rudolph E. Siegel (Basel, Switzerland: Karger, 1976), 3.10.93. See also Rudolph E. Siegel, *Galen on Psychology, Psychopathology, and Function and Diseases of the Nervous System* (Basel, Switzerland: Karger, 1973), 195.

made kings or wolves or demons or birds or artificial instruments."[10] In keeping with this, medical texts of Shakespeare's age uniformly stress the correlation between melancholy and altered cognitive states. For example, Thomas Elyot's *Castel of helth*, a reference work listing beneficial foods and simple medical procedures, includes insomnia and "Dreames fearfull" among the various discomforts attendant on the "Melancholike" complexion.[11] Andrew Boorde's *Breuiary of helthe*, which is focused on diet and other means to physical well-being, describes "melancholy madness" as a "sicknes full of fantasies"; the patient may think himself to "here or to se[e] that thynge that is nat harde nor sene."[12] Philip Barrough's *Methode of phisicke*, a comprehensive guide to ailments ranging from nosebleeds to epilepsy, says that the habitual signs of melancholy are "strange imaginations," such as when "some think them selues brute beastes, and do counterfaite the[ir] voice and noise."[13]

Medical discussions also delineate in intricate detail the precise physiological chain of causation whereby melancholy humor gives rise to these "strange imaginations." Bright, for example, posits that the black bile aggravates the inward wits and that they in turn create the phantasmatic conditions that exacerbate any affective burdens bearing on the heart. He describes how a melancholic vapor released from the spleen may rise into the brain, counterfeiting "terrible obiectes" and "monstrous fictions" in the fantasy. Similar explanations account for the melancholiac's excessive weeping, blushing, and laughing.[14] In a comparable way, André du Laurens says that melancholy swims into the brain by way of the body's circulatory loop: "Spirits and blacke vapours continually passe by the sinewes, veines and arteries, from the braine vnto the eye, which causeth it to see many shadowes and vntrue apparitions in the aire, whereupon from the eye the formes thereof are conueyed vnto the imagination."[15] William Perkins's *Cases of conscience*, invoking the fetid associations we saw in chapter 3, says that corrupted black bile "sends vp noisome fumes as cloudes or mists which doe corrupt the imagination."[16] A still fustier version of this appears in Thomas Nash's *Terrors of the night*: "This slimie melancholy humor still still thickning as it stands still, engendreth many misshapen obiects in our imaginations. . . . Whence haue all these

---

10. Avicenna, *Canon of Medicine*, quoted in *The Nature of Melancholy: From Aristotle to Kristeva*, ed. Jennifer Radden (Oxford: Oxford University Press, 2000), 77.

11. Thomas Elyot, *The castel of helth* (London, 1539), 21.

12. Andrew Boorde, *The Breuiary of helthe* (London, 1547), Aa3r.

13. Philip Barrough, *The methode of phisicke* (London, 1583), 35–37.

14. See Bright, *Treatise of melancholie*, 132–48.

15. André du Laurens, *A discourse of the preseruation of sight*, trans. Richard Surphlet (London, 1599), 92. Earlier published as *Discours de la conservation de la veue* (Tours, 1594).

16. William Perkins, *The first part of The cases of conscience* (London, 1604), 192.

their conglomerate matter but from fuming meteors that arise from the earth, so from the fuming melancholy of our spleene mounteth that hot matter into the higher Region of the braine, whereof manie fearfull visions are framed. Our reason euen like drunken fumes it displaceth and intoxicates, & yeelds vp our intellectiue apprehension to be mocked and trodden vnder foote, by euerie false obiect or counterfeit noyse that comes neere it."[17] With words like "imagination" and "reason," Perkins and Nash invoke the theory of the inward wits. At the same time, with words like "humor," "melancholy," "hot matter," and "fumes," they refer to a paradigm of pathology premised on spirituous fluids and complexions.

The correlation of melancholy and imagination would grow so close as to be inextricable, as Stuart Clark has observed. Chaotic image making prompts melancholia's "primary psychological malfunctions," while melancholy discourse shifts the meaning of *phantasy* from simply "image" to "illusion."[18] Du Laurens writes that the melancholic man is "alwaies disquieted both in bodie and spirit"; "all melancholike persons," he stresses, "have their imagination troubled, for that they deuise with themselues a thousand fantasticall inuentions and obiects, which in deede are not at all."[19] Similarly, Ambroise Paré notes that the primary "Signes of a Melancholicke Person" are "terrible dreames" and nocturnal visions of "Devils, Serpents, darke dens and caves, sepulchers, dead corpses, and many other such things full of horror."[20] In the midst of its long enumeration of the countless signs of melancholia, *The Anatomy of Melancholy* concludes, almost with exasperation, "Who can sufficiently speake of these symptoms, or prescribe rules to comprehend them? . . . If you will describe melancholy, describe a phantasticall conceit, a corrupt imagination, vaine thoughts and different, which who can doe?"[21] The *Anatomy* represents an extreme and perhaps inevitable endpoint, the necessarily florid culmination of a trend that had been gathering since antiquity and progressed with great alacrity in late sixteenth-century medical culture. Merely to scan Burton's table of contents is to understand the reach of melancholy's

---

17. Thomas Nash, *The terrors of the night* (London, 1594), C2v–C3r. Examining Nash's possible influence on Macbeth, Ann Pasternak Slater notes that he uses *melancholy* as a "blanket term" for psychological disturbance. See her "Macbeth and the Terrors of the Night," *Essays in Criticism* 28, no. 2 (1978): 119.

18. Stuart Clark, *Vanities of the Eye: Vision in Early Modern European Culture* (Oxford: Oxford University Press, 2007), 53, 61–63. See as well Winfried Schleiner, *Melancholy, Genius, and Utopia in the Renaissance* (Wiesbaden, Germany: Harrassowitz, 1991), 98–108.

19. Du Laurens, *Discourse*, 82, 87.

20. Ambroise Paré, *The workes of that famous chirurgion Ambroise Parey*, trans. Thomas Johnson (London, 1634), 18. Paré's medical writings were published collectively as *Les oeuvres de M. Ambroise Paré* in 1575.

21. Burton, *Anatomy of Melancholy*, 1:407.

significance; its involvement with devils, stars, education, liberty, and gender; and its diagnostic relevance, as Angus Gowland says, "in a range of intellectual and cultural contexts."[22] Thus, roughly around the time that melancholy was perceived as hopelessly widespread, explorations of its relation to imagination were at their most emphatic and nuanced. Conversely, we could say that the ever more sophisticated explorations of mental imagery in this period involved as a matter of course a consideration of the black bile.

If literary representations are anything to go by, it is clear that melancholic hallucination was an instantly comprehensible trope. Phantastes in *The Faerie Queene* is a figure sorely distracted and "full of melancholy."[23] The protagonist of *The Malcontent*, when asked how he passes his sleepless nights, gleefully exclaims, "I dream the most fantastical," adding, with manic vim, "Dreams, dreams, visions, fantasies, chimeras, imaginations, tricks, conceits!"[24] In *The Two Noble Kinsmen*, the "fancy"-ridden Jailer's daughter is diagnosed with "a most thick and profound melancholy."[25] In *Hamlet* and *Macbeth*, Shakespeare would showcase the trope more centrally; however, the two tragedies interpret the relation between imagination and melancholy differently, in a manner that rather illustrates the sense of culmination that I have been describing. In *Hamlet*, the dramatic intrigue rests on attempts to discern what other people see, or think they see—since, according to the medical logic of the period, the sorts of phantasms one perceives provide clues as to one's general state of health. That these clues can be misread is demonstrated by how easily Hamlet evades his prying observers. But whereas in *Hamlet* fantasy and melancholy are bound by a link that can be manipulated by a skilled actor, in *Macbeth* the link is more difficult to discern. The later play, whose melancholy is not identified explicitly as such, asks in a more open-ended way what exactly imaginative distortion is symptomatic of; it worries over the distinction between pathological and nonpathological perception.

In *Hamlet*, the dark humor is explicitly mentioned twice as an explanation for the prince's temperament. He himself privately muses on "my melancholy"

22. Angus Gowland, *The Worlds of Renaissance Melancholy: Robert Burton in Context* (Cambridge: Cambridge University Press, 2006), 16. For an earlier study, see Lawrence Babb, *The Elizabethan Malady: A Study of Melancholia in English Literature from 1580 to 1642* (East Lansing: Michigan State College Press, 1951).

23. Edmund Spenser, *The Faerie Qveene*, ed. A. C. Hamilton, text ed. Hiroshi Yamashita and Toshiyuki Suzuki (New York: Longman, 2001), 2.9.52.

24. John Marston, *The Malcontent*, in *English Renaissance Drama*, ed. David Bevington, Lars Engle, Katharine Eisaman Maus, and Eric Rasmussen (New York: W. W. Norton, 2008), 1.3.45–56.

25. William Shakespeare, *The Two Noble Kinsmen*, 4.3.44. Shakespeare references are from *The Norton Shakespeare*, ed. Stephen Greenblatt et al. (New York: W. W. Norton, 2016), hereafter cited parenthetically.

(2.2.520), and Claudius speaks of "something in his soul / O'er which his melancholy sits on brood" (3.1.161–62). Understandably, this is the play that readers—among the earliest of which is A. C. Bradley—have tended to associate with Shakespeare's representation of melancholy.[26] In recent years, critics have situated the play in the medical contexts of Shakespeare's time, showing that Hamlet represents a sufferer from melancholy disease as it was understood by physicians such as Bright and Boorde.[27] In particular, Gail Kern Paster's study of the psychophysiology of early modern emotion has established the sophistication and agility of Galenic medicine; Hamlet's volatility, she has shown, may be understood as the inevitable fluctuations of humoral embodiment.[28] Certainly, Hamlet encapsulates the melancholic perturbations listed by Bright: "distrust, doubt, diffidence, or dispaire"; occasional merriness shown in "false laughter"; a diminished ability to "keepe in memory, and to record those thinges, whereof it tooke some custody before."[29] The play's representation of melancholy is also keyed in to adjacent philosophical questions: the relation between body and soul, for example, and skepticism and scholarly melancholy.[30]

We can add to these readings the fact that, according to contemporary imaginative theory, the soundness of Hamlet's psychic state is tied to what others guess to be the phantasms in his mind's eye. Throughout the play, we are watching him closely; in several key scenes, we are also watching others searching him for the symptoms of hallucination. Claudius, Gertrude, Polonius, Rosencrantz and Guildenstern, and Ophelia all attempt to determine Hamlet's motives by judging his demeanor—specifically, by trying to work out the contents of his imagination. Shakespeare invites us too to guess what is going on in Hamlet's head:

HAMLET. My father—methinks I see my father.
HORATIO. Where, my lord?
HAMLET.   In my mind's eye, Horatio. (1.2.184–85)

26. A. C. Bradley suggests that melancholy is the cause of Hamlet's inaction in *Shakespearean Tragedy* (London: Penguin, 1991 [1904]), 120–26. See also Mary O'Sullivan, "Hamlet and Dr. Timothy Bright," *PMLA* 41, no. 3 (1926): 667–79.

27. Carol Falvo Heffernan, *The Melancholy Muse: Chaucer, Shakespeare and Early Medicine* (Pittsburgh: Duquesne University Press, 1995), 123.

28. Gail Kern Paster, *Humoring the Body: Emotions and the Shakespearean Stage* (Chicago: University of Chicago Press, 2004), 45.

29. Bright, *Treatise of melancholie*, 102, 104.

30. Mary Thomas Crane, *Shakespeare's Brain: Reading with Cognitive Theory* (Princeton, NJ: Princeton University Press, 2001), 116–155; Douglas Trevor, *The Poetics of Melancholy in Early Modern England* (Cambridge: Cambridge University Press, 2004), 63–86.

These lines can elicit laughter in performance on account of their irony: Hamlet, who has yet to hear about the nightly sightings of his father's ghost, says he has already "seen" it. A similar moment comes when Hamlet is toying with Polonius, comparing a camel-like cloud to a weasel and then to something "very like a whale" (3.2.348–54). There is a logic to Polonius's cautious questioning; where he fails is in not realizing that the prince suspects as much. *Hamlet* is a play of tests, in which imagination and melancholy are repeatedly presented in the context of a search for proof: the initial need to invoke the apparition of Hamlet Senior; the baiting of Ophelia to decide whether Hamlet is really mad; the theatrical mousetrap that is meant to expose Claudius's crime. Hamlet even examines himself to try and work out if his imagination is being duped:

> The dev'l hath power
> T'assume a pleasing shape; yea, and perhaps
> Out of my weakness and my melancholy,
> As he is very potent with such spirits,
> Abuses me to damn me. (2.2.518–22)

This distinction between the external manipulation of one's "spirits" and melancholic "weakness" was in historical fact a delicate problem, one that complicated discussions of imagination among physicians and demonologists: there was a growing understanding that the supposedly possessed vassals of Satan were oftentimes merely delusional, diseased rather than malicious. Witchhunters such as Lambert Daneau and Henry Holland agreed with skeptics such as Reginald Scot and George Gifford in noting that women claiming to be witches could be sufferers of melancholia and thus could believe that they rode the air or changed shape when they were more likely having vivid hallucinations while lying in their beds.[31]

As a consequence, Renaissance demonological texts routinely reflect on matters of psychology, doing so with a kind of discrimination that surpasses

---

31. On the imagination of witches, see Lambert Daneau, *A dialogue of witches* (London, 1575), K3r; Reginald Scot, *The discouerie of witchcraft* (London, 1584), 58; Henry Holland, *A treatise against witchcraft* (Cambridge, England, 1590), F3r; and George Gifford, *A dialogue concerning witches and witchcraftes* (London, 1593), G2v, K3r. See also Lynn Thorndike, *History of Magic and Experimental Science*, 8 vols. (New York: Columbia University Press, 1966), 8:503–38; Stuart Clark, *Thinking with Demons: The Idea of Witchcraft in Early Modern Europe* (Oxford: Oxford University Press, 1997), 198–210; Katharine Eisaman Maus, "Sorcery and Subjectivity in Early Modern Discourses of Witchcraft," in *Historicism, Psychoanalysis, and Early Modern Culture*, ed. Carla Mazzio and Douglas Trevor (New York: Routledge, 2000), 325–35; and Claudia Swan, *Art, Science, and Witchcraft in Early Modern Holland: Jacques de Gheyn II (1565–1629)* (Cambridge: Cambridge University Press, 2005), chap. 6, 175–94.

the blunt attestations seen in, for example, the *Malleus maleficarum*.[32] Lavater's treatise may be counted among these: his stated aim is to persuade skeptics of the existence of angels and demons by establishing the naturalness of phenomena such as phosphorescence and volcanoes.[33] "Boundary work" of this kind, writes Mary Floyd-Wilson, "sought to distinguish supernatural miracles from preternatural wonders and to provide natural explanations for the apparent powers of non-human matter and agents"; such work, she contends, laid the foundation for "new experimental methods that emphasized the observation of effects over theoretical causation."[34] On the other side of the boundary is, for example, Edward Jorden's tract on the disease known as the "suffocation of the mother." Jorden's hypothesis is stated in his title: "strange actions and passions of the body of man, which in the common opinion are imputed to the Diuell, haue their true naturall causes." Like Lavater, he walks the line between the normal and paranormal, balancing qualification—"I doe not deny but that God doth in these days work extraordinarily"—with cautioning entreaty—"I would in the feare of God aduise men to be very circumspect in pronouncing of a possession."[35] The inexperienced physician, Jorden cautions, must guard himself from his own imagination, lest he wrongly ascribe "straunge accidents" such as the "croaking of Frogges, hissing of Snakes, crowing of Cockes, [or] barking of Dogges" to "some metaphysicall power."[36] He says that treatment and diagnosis should be decoupled: even if a disease were caused supernaturally, its symptoms—convulsions, fits, suffocation—are best tackled with medical rather than ritualistic means.

Alongside scrupulous recommendations regarding diagnostic and interpretive practice, we find taxonomies of atypical mental phenomena such as visions and hallucinations. Lavater's text, for instance, opens with a list of definitions for the terms *spectrum* (a thing that is "seene, eyther truely, or by vaine imagination"); *visum* ("an imagination or a certayne shewe, which men being in sleep, yea and waking also, seeme in their iudgemente to beholde"); and *phantasma* (an appearance of a thing that is not, like "those sightes which

---

32. See the chapters "On the method by which they are transferred in location from place to place" and "The methods by which they change humans into the shapes of wild beasts" in *The Hammer of Witches*, trans. and ed. Christopher S. Mackay (Cambridge: Cambridge University Press, 2009), 2.101A–105C, 2.119A–121A.

33. Lavater, *Of ghostes*, 53.

34. Mary Floyd-Wilson, *Occult Knowledge, Science, and Gender on the Shakespearean Stage* (Cambridge: Cambridge University Press, 2013), 4–5.

35. Edward Jorden, *A briefe discourse of a disease called the suffocation of the mother* (London, 1603), A3r.

36. Jorden, 2r.

men in their sleepe do thinke they see").[37] Daneau's *Dialogue of witches* differentiates between three types of fantastical apparitions: "thinges whiche we haue seene," "thinges whiche our fantasie it selfe hath founde out," and "thinges whyche our imagination conceaueth and deuiseth vppon the wordes and reporte of other."[38] Similarly, Pierre le Loyer's *Treatise of specters* begins by delineating what separates a *phantosme* (a "thing without life") from a *specter* (which "hath a substance hidden and concealed, which seemeth to moue the fantastique body") and other types of vision (literal, imaginative, and intellectual).[39] In Thomas Cooper's thesaurus, *phantasy* is given as a synonym for a remarkable range of words, including *animus* (intellect), *consilium* (judgment), *ratio* (reason), *simulacrum* (likeness), *stomachus* (ill temper), and *visum* (vision).[40] Although the categorical distinctions made by early modern cognitive theorists are not consistent with one another or even especially crisp, it is evident that such thinkers were keen to map the full spectrum of cognitive dysfunction in all its complexity.

If we did not at least hypothetically believe it possible to discern the particulars of Hamlet's disposition just by looking at him, then Polonius's failure to do so would be rather less entertaining. The Danish play hinges on the knowledge that the art of diagnosis is a demanding one. Not only are phantasms shifting and insubstantial things, our sense of other people's minds is woefully conjectural. The metatheatricality of *Hamlet*, especially the performative talents of its protagonist, drive this point home, for if there is certain proof that we cannot know what goes on inside others, it lies surely in the chimerical actor who demonstrates how two sensibilities—his own and that of the phantom being whom he portrays—may inhabit a single mind. Beneath the playful morphing of camels into whales, then, lies a more serious acknowledgment of the difficulty of distinguishing between what the melancholic sees and what he thinks he sees—and doing so, moreover, from an external vantage point.

Hamlet demonstrates this difficulty by multiplying the observational viewpoints, presenting the same reality from different perspectives. Consider, for

37. Lavater, *Of ghostes*, 1–2.

38. Daneau, *Dialogue of witches*, G3r–v.

39. Pierre le Loyer, *A treatise of specters or straunge sights*, trans. Zachary Jones (London, 1605), 2. Le Loyer's earlier demonological titles include *IIII. livres des spectres* (Angers, 1586) and *Discours des spectres ou visions* (Paris, 1586).

40. Thomas Cooper, *Thesaurus linguae Romanae & Britannicae* (London, 1578). Some examples: *Obsequi animo*, "To follow hys phantasie" (I1r); *Consilium sibi capere*, "To do after his owne phantasie" (Q3v); *Mea est sic ratio*, "This is my fantasie" (Oooo6r); *Simulachra inania*, "vaine phantasies" (Ppp6v); *Stomachus*, "minde: fantasie" (Ddddd5r); *Visum*, "a dreame: a fantasie" (Qqqqqq4v).

instance, Ophelia's description of the prince soon after his encounter with the ghost:

> He falls to such perusal of my face
> As 'a would draw it. Long stayed he so;
> At last, a little shaking of mine arm
> And thrice his head thus waving up and down,
> He raised a sigh so piteous and profound
> As it did seem to shatter all his bulk
> And end his being. That done, he lets me go,
> And with his head over his shoulder turned
> He seemed to find his way without his eyes,
> For out o'doors he went without their helps,
> And to the last bended their light on me. (2.1.87–97)

Why are we provided with this description when we just now directly saw Hamlet speaking with the ghost? Ophelia draws our attention to the prince's emotional state—"a sigh so piteous and profound"—but she is also struck by his appearance, a spectacle in itself. In her account, Hamlet's sight weaves between the literal and imaginary: "He falls to such perusal of my face"; "He seemed to find his way without his eyes." Galenic physiognomy held that it is possible to discern medical symptoms through facial observation, but here it seems that the invisible object of Hamlet's gaze is of as much interest as his countenance.[41]

"Whereon do you look?" asks Gertrude in the closet scene. The reappearance of the ghost and its effect on Hamlet drive this scene's phantasmatic energy, but so too does the queen's alternating inward and outward gaze. Hamlet, determined to expose what he sees as his mother's moral blindness, thrusts pictures at her: "This was your husband. Look you now what follows" (3.4.63). The images eventually force her to look within:

> O Hamlet, speak no more!
> Thou turn'st my very eyes into my soul,
> And there I see such black and grievèd spots
> As will leave there their tint. (3.4.88–91)

The dark "tint" mentioned evokes the black bile, as though just by manipulating Gertrude's inner eye Hamlet has engendered a melancholic fume in his

---

41. Siegel, *Galen on Psychology*, 202.

mother. When the ghost suddenly appears, the perspectival arrangement shifts. Now Gertrude is observing Hamlet:

> Alas, how is't with you
> That you do bend your eye on vacancy
> And with th'incorporal air do hold discourse?
> Forth at your eyes your spirits wildly peep,
> And as the sleeping soldiers in th'alarm
> Your bedded hair like life in excrements
> Start up and stand on end. O gentle son,
> Upon the heat and flame of thy distemper
> Sprinkle cool patience. Whereon do you look? (3.4.115–23)

We might again ask, why does Gertrude describe her son at this moment, when we see him too? She calls attention to the invisible and immaterial: the "spirits" peeping in Hamlet's wild eyes; the "alarm" that rouses the "sleeping soldiers" of his hair; the internal "heat and flame" of his "distemper." The point, it seems, is to draw attention precisely to what we cannot see or feel: the feeling that ripples across his skin, raises the temperature of his blood. This is interiority represented from outside, as a set of secrets never to be laid bare, and for a moment we register the insuperable distance that remains between us and Shakespeare's most eloquent protagonist.

> GERTRUDE. This is the very coinage of your brain:
>   This bodiless creation ecstasy
>   Is very cunning in. (3.4.138–40)

There are two phantom entities in the room: the ghost, and Hamlet's own mental "ecstasy," described by Gertrude as a "cunning" thing in itself. "It is not madness / That I have uttered," retorts the prince. "Bring me to the test" (3.4.142–43).

Thus, *Hamlet*'s representation of melancholy tests the idea that the phantasmatic imagination provides a means of penetrating the mind's interior. In practice it is almost impossible to do this, the play shows—not just with its metatheatrical reminders of the illusoriness of performance but also by repeatedly challenging us to question and monitor Hamlet's inward and literal vision. It is not only Hamlet's fantasy that is of interest; we are also called on to surmise what Claudius sees in the "The Murder of Gonzago," to acknowledge the dark spots in Gertrude's soul, and to guess at the nameless actors of mad Ophelia's bawdy lyrics. The play's tangled sightlines—its apparitions, hallucinations, introspections, and scrutinizing stares—dramatize the problem of making judgments about the trustworthiness of another person's judgment on the basis of his or her perceptions. *Hamlet* thus subscribes to the ties between mental imagery and mental health widely accepted in sixteenth-century discourse, albeit conveying the difficulty of discerning those connections in practice.

Macbeth is a continuation of these themes, but its treatment of them is more pointedly ambiguous. With its pervasive hallucinatory intensity, the Scottish play is a more comprehensive and yet more abstract representation of melancholy. Whereas Hamlet's melancholy is more individual and affective, roiling inside him, the melancholy of *Macbeth*, if we may call it that, is impersonal and perceptual, residing in the supposed boundary that separates the inner and the outer. Whereas in *Hamlet* the diagnostic link between melancholy and hallucination is open to construal, *Macbeth* obscures that link by never mentioning melancholy explicitly. The point is not so much to decide whether Macbeth is melancholic as to ask what it is about the hallucinatory ecstasies that normally go with melancholy that qualifies as disorder. We are no longer speaking only of black bile, *Macbeth* seems to say. There are other questions being negotiated under the sign of melancholy, as it were, to do with the distinction between illness and ordinary human impulses, between figments in the mind and things in the world.

A Modern Ecstasy

The importance of imagination has been much noted in the critical discourse of *Macbeth*. Readers have noticed that the play's imagery is startlingly vivid, its epistemology of seeing particularly vexed.[42] Its dramaturgy toys with the viewer's imagination, casting us as Macbeth's "visionary accomplices."[43] Often stressed is Macbeth's own extraordinary power of fantasy, which may be his essential flaw: he is a "prisoner of his own imagination, bound into doubts and fear." Both he and Lady Macbeth, indeed, may be said to employ a sort of visual thinking.[44] To these readings we may fruitfully add a specialized

---

42. See Kenneth Muir, "Image and Symbol in Macbeth," *Shakespeare Survey* 19 (1966): 45–54; Richard S. Ide, "Theatre of the Mind: An Essay on Macbeth," *ELH* 42, no. 3 (1975): 338–61; D. J. Palmer, "'A New Gorgon': Visual Effects in Macbeth," in *Focus on "Macbeth,"* ed. John Russell Brown (London: Routledge, 1982), 54–69; Lucy Gent, "The Self-Cozening Eye," *Review of English Studies* 34, no. 136 (1983): 419–28; and Houston Diehl, "Horrid Image, Sorry Sight, Fatal Vision: The Visual Rhetoric of *Macbeth*," *Shakespeare Studies* 16 (1983): 191–203.

43. Thomas Cartelli, "Banquo's Ghost: The Shared Vision," *Theatre Journal* 35, no. 3 (1983): 393.

44. See, respectively, R. A. Foakes, "Images of Death: Ambition in Macbeth," in Brown, *Focus on "Macbeth,"* 27; Arthur F. Kinney, *Lies like Truth: Shakespeare, Macbeth, and the Cultural Moment* (Detroit: Wayne State University Press, 2001), 268–72.

consideration of mental representation as it was understood in the early modern period. The play itself alludes to faculty psychology. When, for example, Lady Macbeth decides to drug the chamberlains such that "memory, the warder of the brain, / Shall be a fume, and the receipt of reason / A limbeck only" (1.7.65–67), she is likening the ventricles of the brain to vessels of liquid distillation; reason, she is saying, can be clouded by vapors of memory. Likewise, when she rebukes Macbeth for indulging his mind's "sorriest fancies," for "using those thoughts which should indeed have died" (3.2.9–10), she accuses him of clinging to phantasms that ought to have been forgotten.

Psychic distress threatens to undo them both. On the night of Duncan's murder, Lady Macbeth warns her husband not to "unbend your noble strength" by thinking "So brainsickly of things" (2.2.48–49). He in turn tries to clear the "perilous stuff " that "weighs upon [her] heart" (5.3.44–45). Life is described as a "fitful fever" (3.2.23)—a complex ailment, according to early modern physic. Pressing on the people of this play is a malady masked by vague designations such as "disease," "affliction," and "fit." The diagnosis that may well have struck Jacobean audiences is melancholy. Critics have not much tended to associate melancholy with *Macbeth*, even though many of the play's phantasmatic events—the vanishing witches, the floating dagger, the ghost at the banquet table, the pageant of ghostly kings—chime with contemporary melancholic discourse. In fact, troubled sleep, visual and auditory hallucinations, and a general sense of terror were known symptoms of melancholy.[45] So too, the "present fears" that unfix Macbeth's hair echo the belief that melancholy can "cause the interiour spirits of the braine to waxe verie *wilde* and *fearefull*."[46] The paranoia that leads him to slaughter Macduff's family was a signature of the melancholiac: "His phantasia workes, and he imagineth, that the thing is alreadie, or shall befall him"; he is ever "suspitious" of "a matter of farther feare."[47] Melancholy men, doctors say, will fantasize of becoming a "King, an Emperour, a Monarke."[48]

Certain events from the play even echo anecdotes told in Lavater's treatise on specters. In one, a murderer sees his victim in a fish's head on his dinner plate—recalling the return of Banquo at the banquet table. Another tells of an outnumbered military leader alarmed to see an approaching mass of pikes

45. See Bright, *Treatise of melancholie*, 195, which speaks of "vaine feares, and false conceits of apparitions, imagination of a voyce sounding in your eares, frightfull dreames."

46. John Deacon, *Dialogicall discourses of spirits and divels* (London, 1601), 160.

47. Perkins, *Cases of conscience*, 194; Bright, *Treatise of melancholie*, 133.

48. Du Laurens, *Discourse*, 98. See also Bright, *Treatise of melancholie*, 133: "Their estate, seeminge in their owne fantasie much worse then it is, or then the condition of other men, maketh them desire that they see other to enjoy."

and spears, only to realize he is looking at an overgrowth of shrubs—the moving forest of Birnam.[49] Of course, the forest in *Macbeth* is a real army; some of the play's most terrifying scenes are not hallucinated at all. Though it would have been possible to imagine the wayward sisters as melancholic, their occult power could just as easily be taken as authentic. And it is hard to decide if Macbeth is melancholic, given how little we see of him before he meets the witches—he has only two lines before the prophecies are spoken. The point is that, despite the play's oppressive fantastical aspect, it avoids interrogating cognitive phenomena. Macbeth is quick to call the floating dagger a figment of his brain; the physician avoids treating Lady Macbeth for what he thinks is a spiritual illness. Whereas the ghost of *Hamlet* makes demands and issues instructions, communication with otherworldly realms is abortive in *Macbeth*. Invocations are made unto air; ghosts resist conversation. Just as its melancholy remains unmarked, this play's hallucinations are not easily interpreted as such.

A sense of interiority, meanwhile, is elusive. Powerful fantasies are often being had, but it is not always clear to whom they belong. Upon first meeting the three hags, Banquo asks,

Are ye fantastical or that indeed
Which outwardly ye show? (1.3.54–55)

The two possible meanings of "fantastical" sum up the perceived ambiguity surrounding the agency of witches: the word refers to something fantastic or unreal, as well as to a person who is prone to fantasy. While Banquo is questioning the women, something strange comes over Macbeth—he becomes "rapt withal," himself a transfixed "fantastical." "Smothered in surmise," he asks himself,

Why do I yield to that suggestion
Whose horrid image doth unfix my hair
And make my seated heart knock at my ribs
Against the use of nature? (1.3.136–39)

Banquo cannot help but repeat his earlier observation: "Look how our partner's rapt" (1.3.145). Both the Macbeths are transformed by the somatic effects of imagination. Lady Macbeth's invitation to spirits to "unsex" her, for example, is a plea for a kind of physical reconstitution: "Fill me from the crown to the toe"; "make thick my blood"; "take my milk" (1.5.38–46). In both him

---

49. Lavater's text was available in English before the publication of Raphael Holinshed's *Chronicles*, which also describes a moving forest.

and her, there is less of a sense of Hamlet's churning interior than of the uncanny numbness of the automaton. They give up their bodies as vessels to foreign agents; they walk and talk while asleep; their seeing eyes "wink" at the bloody doings of their hands (1.4.52).

"What is happening in me?" Hamlet seems to wonder, pinned between his own psychological fragility and the supernatural manifestations at Elsinore. What has happened to the world is the implicit anxiety of *Macbeth*, whose protagonists murder sleep, incarnadine the ocean, and raise the dead from their graves. "Let the frame of things disjoint," says Macbeth:

> Better be with the dead,
> Whom we, to gain our peace, have sent to peace,
> Than on the torture of the mind to lie
> In restless ecstasy. (3.2.19–22)

It is not only Macbeth who is gripped by "ecstasy." In the play's second scene, the Thane of Ross, coming from battle, is enraptured:

> What a haste looks through his eyes!
> So should he look that seems to speak things strange. (1.2.46–47)

When, on the morning of the murder, the chamberlains awake, we are told "they stared and were distracted" (2.3.101). Duncan says of the executed Cawdor, "There's no art / To find the mind's construction in the face (1.4.11–12). Sean H. McDowell notes rightly that the play alludes to the notion that it is possible to conjecture in the body "a person's inward disposition."[50] Yet, in *Macbeth*, such conjectures are thwarted as often as they are invited. In *Hamlet*, homicidal guilt or lovesickness can be discerned through careful scrutiny; in *Macbeth*, the enthralled mind is too often spirited away from its bodily home.

This rapture seems moreover to be endemic rather than isolated, inescapable rather than occasional, as is confirmed in Ross's lament for ailing Scotland:

> Alas, poor country,
> Almost afraid to know itself. . . .
> . . . .
> Where sighs and groans and shrieks that rend the air

---

50. Sean H. McDowell, "Macbeth and the Perils of Conjecture," in *Knowing Shakespeare: Senses, Embodiment and Cognition*, ed. Lowell Gallagher and Shankar Raman (Basingstoke, England: Palgrave Macmillan, 2010), 36.

Are made, not marked; where violent sorrow seems
A modern ecstasy. The dead man's knell
Is there scarce asked for who, and good men's lives
Expire before the flowers in their caps,
Dying or ere they sicken. (4.3.164–73)

Ecstasy, which means literally to stand outside oneself, renders unrecogniz-
able what should be familiar.[51] The bells for the dead go unheeded, groans
of suffering are "not marked," death precedes sickness. Ross's vision extrapolates Macbeth's personal "ecstasy," the self-alienation of a single mind, into a
universal condition that is "modern" in the sense of being commonplace as
well as current.[52] Here is not a complete breakdown of social relations. The
knells are still tolled; "sorrow" indicates a residue of humane feeling; "good
men" do exist, if only to die. Still, the "ecstasy" is disturbingly faceless. Pain
and suffering have become unconnected from individuals, leaving the country awash with a distress that belongs to everyone and no one. As Clark says
in his reading, Macbeth underlines the way that perceptual unreliability can
become a political issue, "an accompaniment to political treason and moral
turmoil."[53]

The idea of ecstasy surfaces in Neoplatonist accounts of fascination, sympathy, and inspiration. Aristotle had said that sibyls and soothsayers exhibit the
frenzy of a brain overheated by melancholy.[54] In the Renaissance, the connection between melancholia and Platonic furor was theorized by Marsilio Ficino, who delineated the ties between frenzy, genius, and the intellectual
temperament.[55] Ross's collective ecstasy also evokes Pietro Pomponazzi, who

---

51. See La Primaudaye, *Second part of the French academie*, 165, which deems "such as are op-
pressed with melancholy" to be "beside themselues." Wilhelm Adolf Scribonius describes an "Exsta-
sie" to be a "vehement imagination of the departure (for a time) of the soule from the bodie" in his
Naturall philosophy, trans. Daniel Widdowes (London, 1621), 51–52.

52. *OED Online*, s.v. "modern, adj. and n." (Oxford: Oxford University Press, January 2018).
http://www.oed.com/view/Entry/120618. See sense 4, "everyday, ordinary, commonplace," and
sense 2a, "of or relating to the present and recent times."

53. Clark, *Vanities of the Eye*, 254.

54. Aristotle, *Problems*, in *Problems, Volume II: Books 20–38. Rhetoric to Alexander*, trans. and ed. Robert Mayhew and David C. Mirhady, Loeb Classical Library 317 (Cambridge, MA: Harvard University
Press, 2011), 954a34–b5.

55. Marsilio Ficino, *Three Books on Life*, trans. and ed. Carol V. Kaske and John R. Clark (Binghamton, NY: Medieval & Renaissance Texts & Studies in conjunction with the Renaissance Society of
America, 1989), 117–23. Ficino writes that melancholy "terrifies the soul, and dulls the intelligence,"
yet, when properly tempered, it enables the mind to explore eagerly, with sound judgment and longer
retention (117, 121). See also Noel L. Brann, *The Debate over the Origin of Genius during the Italian Renaissance* (Leiden, Netherlands: Brill, 2002), 82–107; and Guido Giglioni, "Coping with Inner and
Outer Demons: Marsilio Ficino's Theory of the Imagination," in *Diseases of the Imagination and*

maintained that, when performed before credulous and receptive minds, imaginative acts could alter the flow of spirits in the air and so form strange visions.[56] All this being said, though, the imaginings of Macbeth could hardly be mistaken for instances of divine furor. The play has little interest in Ficinian transcendence, with the soul rising closer to God; its malaise seems more to do with not knowing oneself than with self-knowledge. And unlike the collective hallucinations of Pomponazzi, Scotland's ecstasy is burdened with real and measurable loss.

A repeating trope in sixteenth-century medical discourse takes the human body as a sort of fortress—the "castle of health"—besieged from both within and without, by humoral imbalances as much as contagious miasmas.[57] The presumed wall dividing the interior from the exterior, what Stephen Greenblatt calls the membrane between imagination and the world, is inscribed in the very paradigm of faculty psychology—which is a psychology of "inward" wits as opposed to "outward" senses.[58] The devil's power, ran the popular view, is that he can straddle the gap, create a *"commotion of humours"* as easily in "the *interiour,* as *exterior senses.*"[59] As Le Loyer says, the adversary insinuates himself "into the inward and interior senses, and into the fantasie of men," which he moves "in the same sorte as hee dooth the externall," causing "a certayne extasie and alienation of their spirites."[60] Intriguing as these explanations are, the supposed limen between the inward and the outward is wholly theoretical; it is a border that no anatomist would ever be able to find. But, having creating it, the discourse must keep affirming it, schematizing it with elaborate classifications that enforce the separation between the two realms. Staved off is the less intuitive notion that the mind encompasses elements both internal and external, the notion that our cognitive selves inevitably shape, and become shaped by, things that we superficially perceive as existing outside us.

Imaginary Disease in the Early Modern Period, ed. Yasmin Annabel Haskell (Turnhout, Belgium: Brepols, 2011), 19–51.

56. Pomponazzi suggested that many apparently paranormal phenomena were either illusions crafted by priests or produced by natural properties of air. See Martin L. Pine, *Pietro Pomponazzi: Radical Philosopher of the Renaissance* (Padua, Italy: Editrice Antenore, 1986), 244–45; Daniel Pickering Walker, *Spiritual and Demonic Magic: From Ficino to Campanella* (University Park: Pennsylvania State University Press, 2000), 107–8; and Thorndike, *History of Magic,* 5:97, 106–7.

57. See Jean Fernel, *Jean Fernel's "On the Hidden Causes of Things": Forms, Souls, and Occult Diseases in Renaissance Medicine,* ed. John M. Forrester (Leiden, Netherlands: Brill, 2005), 43.

58. Stephen Greenblatt, "Shakespeare Bewitched," in *New Historical Literary Study: Essays on Reproducing Texts, Representing History,* ed. Jeffrey N. Cox and Larry J. Reynolds (Princeton, NJ: Princeton University Press, 1993), 123.

59. Deacon, *Dialogicall discourses,* 143.

60. Le Loyer, *Treatise of specters,* 123v.

Macbeth is premised on the idea that, as Paster puts it, the body is a change-able "internal microclimate" that is porously receptive to the outside world.[61] Yet the intrigue of the play lies in its suggestion that the premise of interiority may in itself be an occlusive fiction. Lady Macbeth's sleepwalking, for exam-ple, unsettles the premise of a privately confined mind, foregrounding instead the way the psyche extends beyond its bodily vessel. In the scene, her speeches drift forward and backward in time, addressed to various others as well as the self: "Out, damned spot! Out, I say! One, two, why, then, 'tis time to do't. Hell is murky. Fie, my lord, fie, a soldier and afeard? What need we fear? Who knows it when none can call our power to account? Yet who would have thought the old man to have had so much blood in him?" (5.1.31–35). In a single utterance, we move from the spot of blood on Lady Macbeth's hand, discovered after the murder, to the imperatives she issued to her husband beforehand, and then to the eternity of damnation—"Hell is murky." It continues in this way: "I tell you yet again, Banquo's buried"; "To bed, to bed. There's knocking at the gate" (5.1.56, 59). Though the lines are lurching and incoherent, their sentiments are not unfamiliar: Lady Macbeth says almost nothing here that we have not heard her express before. Something is not right—but what? In *Hamlet*, we see mad Ophelia in the conventional onstage attitude of female unraveling. Her hair is down; she is strumming a lute, singing haunting ditties. Yet, although she is crazed with grief, she still recognizes the other characters on stage, gifts them with imaginary flowers. What could be the cognitive explanation of what is happening in Lady Macbeth's mind? We hear that she receives the "benefit of sleep" while "do[ing] the effects of watching" (5.1.9). Her eyes are open but "their sense are shut" (5.1.22). Is she having a bad dream or recalling past events? What, and how, does she "see"? Lady Macbeth's condition would likely have fascinated early modern psychologists. This extraordinary episode supports MacDonald's idea that such dramatic representations of madness were a way of pondering the nature of mental illness.[62]

Not only is this episode framed in a more clinical context than Ophelia's poetic dissolution, it is presented, specifically, as a diagnostic problem: with the observing doctor onstage, we are invited to consider the case as he does.[63] We notice the mediating role of the medical profession—the physician stands

---

61. Paster, *Humoring the Body*, 19.

62. MacDonald, in *Mystical Bedlam*, 121–22, strikingly implies that the incidence of madness on the English stage was disproportionate to the reality, given the relative smallness of London's Beth-lem Hospital.

63. Ophelia's distraction has also been read in clinical terms, of course: see Carol Thomas Neely, *Distracted Subjects: Madness and Gender in Shakespeare and Early Modern Culture* (Ithaca, NY: Cornell University Press, 2004), 52–53, which describes how Ophelia's madness builds on existing ideas about erotomania but is innovative too.

directly between the patient and our understanding of her. Observing the observers, we become aware of their exterior stance, as well as ours. At the
same time, the dramaturgy overturns the idea that inwardness is a type of con-
tainer: we realize, uncomfortably, that the turmoil enclosed within Lady
Macbeth is also inscribed invisibly in the room around her, though we cannot
see it. The usual relation of exchange between body and environment—which
controls what "crosses the threshold of the body, from within or without"—
is, in a way, broken down by the phenomenon of hallucination.[64] As a result,
the end diagnosis falls a little flat: the doctor pronounces the matter to be more
suited to "the divine than the physician" (5.1.67). Objectively, his deduction is
legitimate: as Bright explained, pangs of conscience are not melancholy; they
fall outside the purview of medicine.[65] Yet, as Christopher Tilmouth has interestingly suggested, conscience itself is an ecstatic attitude, a pose of exteri-
orized self-scrutiny in which "the subject suddenly becom[es] a detached
viewer of himself."[66] We might say, then, that Lady Macbeth has become unhinged not just by guilt but by an ecstasy of guilt, by a part of herself detach-
ing from herself. But her doctor does not say any of this: instead, he speaks
cryptically of "foul whisp'rings . . . abroad" and "infected minds" (5.1.64–65),
evoking Ross's "modern ecstasy." It may be that he feels there is nothing
uniquely the matter with Lady Macbeth, and he must in any event negotiate
carefully the complexities of treating a royal patient, one whose hands may
be figuratively if not literally bloody.[67] Still, the simplistic outcome—the falsely
confident distinction between spiritual and medical malady—makes a point
in itself. It implies the impossible position of early modern psychopathology,
which was committed to making fine categorical discriminations even as it was
combating a medical calamity whose nature and reach seemed wildly indiscriminate.

In his subsequent interview with the king, the doctor says nothing of summoning a "divine." He notes instead that the queen is

---

64. Mary Floyd-Wilson and Garrett A. Sullivan Jr., introduction to *Environment and Embodiment in
Early Modern England*, ed. Mary Floyd-Wilson and Garrett A. Sullivan Jr. (Basingstoke, England: Palgrave Macmillan, 2007), 4.

65. Bright, *Treatise of melancholie*, 189.

66. Christopher Tilmouth, "Shakespeare's Open Consciences," *Renaissance Studies* 23, no. 4
(2009): 503, 511. On conscience, see too Frederick Kiefer, "'Written Troubles of the Brain': Lady Macbeth's Conscience," in *Reading and Writing in Shakespeare*, ed. David M. Bergeron (Newark: University
of Delaware Press, 1996), 64–81; and Abraham Stoll, "Macbeth's Equivocal Conscience," in *Macbeth:
New Critical Essays*, ed. Nick Moschovakis (New York: Routledge, 2008), 132–50.

67. These suggestions are made respectively by David Hoeniger, *Medicine and Shakespeare in the
English Renaissance* (Newark: University of Delaware Press, 1992), 65, and Todd H. J. Pettigrew, *Shakespeare and the Practice of Physic: Medical Narratives on the Early Modern English Stage* (Newark: University
of Delaware Press, 2007), 66.

> not so sick, my lord,
> As she is troubled with thick-coming fancies
> That keep her from her rest. (5.3.37–39)

Macbeth responds,

> Cure her of that.
> Canst thou not minister to a mind diseased,
> Pluck from the memory a rooted sorrow,
> Raze out the written troubles of the brain,
> And with some sweet oblivious antidote
> Cleanse the stuffed bosom of that perilous stuff
> Which weighs upon the heart? (5.3.39–45)

The physician, perhaps sensing that they are no longer talking about Lady Macbeth, switches to the male pronoun "Therein the patient / Must minister to himself" (5.3.45–46). Though he echoes Macbeth's "minister," a word that implies both medical and spiritual dispensation, he is again laconic. Here are none of the effusive consolations provided by Bright, for instance, who assures the despairing melancholic that the fantasy of hopelessness will pass—"giue no credite thereunto, but as it is, so esteeme it a delusion which time will discouer and lay open."[68] Macbeth goes on:

> If thou couldst, Doctor, cast
> The water of my land, find her disease,
> And purge it to a sound and pristine health,
> I would applaud thee to the very echo
> That should applaud again. (5.3.50–54)

The irony of these lines is sharp, given this play's representation of mental illness. The cures that Macbeth suggests—"rhubarb, cyme, or some purgative drug"—are somewhat inapposite. Historically, fancies were not easily eliminated by purgative drugs; rather, fire had to be fought with fire. Galen had found sympathetic trickery effective, and Jorden suggests something similar for those resistant to "reason and perswasions": "We may politikely confirme them in their fantasies, that wee may the better fasten some cure vpon them."[69]

---

68. Bright, *Treatise of melancholie*, 222.
69. See Winfried Schleiner, *Medical Ethics in the Renaissance* (Washington, DC: Georgetown University Press, 1995), 22; Jorden, *Suffocation of the mother*, 24v.

It may be that Macbeth full well knows that sorrow cannot so easily be extracted, oblivion so easily induced. His lines express a desperate wish—a very modern-seeming wish—to medicalize mental anguish such that we might at least hope to "raze" it with scientific means. The passage also registers the collectivized distress intoned throughout the play. In fact, Scotland is indeed about to be purged, though not by Macbeth: Malcolm, "med'cine of the sickly weal," is mobilizing his forces (5.2.27). There is a blind spot in Macbeth's characterization of diseased Scotland, which conveniently diminishes the role that he himself has played in its creation. Medicine does not emerge as a means of resolution because etiologies of the pathological imagination are mixed up with the basic operations of human volition. Even if Macbeth refuses to confront this truth fully, Shakespeare ensures that we see it.

Hideous Phantasma

Early modern accounts of imagination cast the fantasy as a creature at odds with the rest of one's person, wayward and uncontrollable. Pierre Charron's *Of wisdome* says that an inflamed imagination is "actiue and stirring," so much so that it dominates the other faculties; it presents delusions, "if it will, to the vnderstanding," or else, "if it will, it commits them to the memorie."[70] Similarly, Le Loyer notes that imagination likes to "accommodate her selfe," to "imprint in it selfe many things by reason of Maladies, Fevers, Melancholics, Doatings, and Frensies," "to tickle and delight it selfe with such idle conceipts as shall be most pleasing and agreeable vnto it selfe."[71] A kind of self-serving malevolence, the rogue fantasy raises the troubling idea that, suspended as we are between the tyranny of our mortal shells and the pitiless wars of angels, our thoughts are not our own. Whether the source of the illness is the body, the mind, or some external influence, the imperative remains to guard the self from alien agents.

Yet, often missing in the medical discourse of imagination is the kind of hallucination that has nothing to do with demons and hags—the kind that features in the play's second act.

Is this a dagger which I see before me,
The handle toward my hand? Come, let me clutch thee.
I have thee not, and yet I see thee still.

---

70. Pierre Charron, *Of wisdome*, trans. Samson Lennard (London, 1608), 50. Earlier published as *De la Sagesse* (Bordeaux, 1601).
71. Le Loyer, *Treatise of specters*, 96v.

Art thou not, fatal vision, sensible
To feeling as to sight? Or art thou but
A dagger of the mind, a false creation
Proceeding from the heat-oppressèd brain?
I see thee yet in form as palpable
As this which now I draw.
Thou marshall'st me the way that I was going,
And such an instrument I was to use.
Mine eyes are made the fools o'th' other senses,
Or else worth all the rest. I see thee still,
And on thy blade and dudgeon gouts of blood,
Which was not so before. There's no such thing.
It is the bloody business which informs
Thus to mine eyes. (2.1.33–49)

Even as Macbeth asks the question, "Is this a dagger," he is already grasping at its handle: "Let me clutch thee." He barely hesitates in deciding that the hallucination is pathological and not preternatural, born of a qualitative imbalance in his head. This crucial decision, which would have given Shakespeare's more medically and religiously minded contemporaries some pause—in *Hamlet* it took four men to verify the existence of the ghost—Macbeth makes in an instant. He speaks to the dagger: "Art thou not . . . sensible." "Sensible" can mean "perceptible," as well as "capable of sensation"—as though the dagger were itself a thinking thing. These are not cautious addresses unto the supernatural realm, the words of a man glancing furtively about for diabolic antagonists; this is a man in dialogue with a projection of himself.

Macbeth registers the dagger no fewer than three times: "yet I see thee still"; "I see thee yet"; "I see thee still." The triple chiming of "see thee" protracts the vision. Time seems to slow down. Then, with "there's no such thing," he almost wills the illusion away, revising its cause. It is no longer a perceptual error; it is "the bloody business" at hand. "Inform" subtly echoes "in form" from an earlier line, hinting at the inward formation of phantasms. "Bloody business," however, is disingenuous, skipping lightly over murder. The questions are not really questions; the second-person pronouns convey familiarity rather than alarm; the explanations offered are a bit too assured. And the speech does not end when the dagger disappears. Rather, it takes a turn:

Now o'er the one half world
Nature seems dead, and wicked dreams abuse
The curtained sleep. Witchcraft celebrates

Pale Hecate's off'rings, and withered Murder,
Alarumed by his sentinel the wolf,
Whose howl's his watch, thus with his stealthy pace,
With Tarquin's ravishing strides, towards his design
Moves like a ghost. (2.1.49–56)

Gone is the first-person voice. We now have a ghoulish nocturne populated with "dreams," "witchcraft," and "murder." Pagan mythology collides with folklore; nature mingles with the afterworld. How have we come from "there's no such thing" to this mythical landscape of ghosts and wolves? It is as though Macbeth is deliberately crowding his mind, burying the dagger with conjured phantoms; these patently unreal, consciously fashioned phantasms seem designed to make us question the provenance of the initial dagger illusion.

The dagger apparition is the materialization as a mental image of an intention: it stands for the decision to kill Duncan. The decision is not articulated in words, nor is it a purely abstract thought. Instead, it is a vision, a partial visualization of an event—an event that is heralded, made imminent, by that very visualization. A similar moment arises earlier in the play:

Present fears
Are less than horrible imaginings.
My thought, whose murder yet is but fantastical,
Shakes so my single state of man
That function is smothered in surmise,
And nothing is but what is not. (1.3.139–44)

Here, future imaginings dominate present fears. Bodily "function" is subjugated by "surmise," and "nothing is but what is not." The force that shakes Macbeth's "state" is not yet a fully formed "thought"; it is "but fantastical." Yet its nigh-prophetic potency is extreme. *Macbeth* is not the first play in which Shakespeare links the surmising of a terrible crime to the imaginative faculty. Compare the foregoing speech with this one from *Julius Caesar*:

Between the acting of a dreadful thing
And the first motion, all the interim is
Like a phantasma or a hideous dream:
The genius and the mortal instruments
Are then in council, and the state of man,

Like to a little kingdom, suffers then
The nature of an insurrection. (2.1.63–69)

These lines, spoken by Brutus, might as well have come from the early acts of
Macbeth. It is the only time Shakespeare uses the word *phantasma*, and he does
so to describe the enveloping haze of intention. Brutus and Macbeth use the
same metaphorical vehicle: to conceive of political insurrection induces insurrection within the "little kingdom" of the self. G. Wilson Knight finds in both
speeches a "sickly sense of nightmare unreality," "a black abyss of nothing."[72]
Yet, if anything, imagining invokes too much rather than too little feeling. Both
mind and body, "genius" and "mortal instruments," are wholly taken over by
the "hideous dream."

There are other instances in Shakespeare where imagination and political de-
signs go together. In *1 Henry IV*, the Earl of Northumberland notes of Hotspur,

Imagination of some great exploit
Drives him beyond the bounds of patience. (1.3.198–99)

In *2 Henry IV*, Bardolph says, again of Hotspur, that he,

with great imagination
Proper to madmen, led his powers to death,
And, winking, leapt into destruction. (1.3.27–29)

In *The Tempest*, Sebastian says to Antonio,

My strong imagination sees a crown
Dropping upon thy head. (2.1.201–2)

And Hamlet confesses, "I am very proud, revengeful, ambitious, with more
offenses at my beck than I have thoughts to put them in, imagination to give
them shape, or time to act them in" (3.1.122–25).[73] In a way, a tragedy of king
killing is the generic form perfect for depicting the hallucinatory phantasma
that forms the bridge between inception and execution. As Christopher Pye

---

72. G. Wilson Knight, *The Wheel of Fire* (London: Routledge, 2001), 138.
73. James Calderwood, in *If It Were Done: "Macbeth" and Tragic Action* (Amherst: University of
Massachusetts Press, 1986), 3, 7, suggests that Hamlet's placement of "imagination" in between
"thought" and "action" is significant: for him, "imagination is an impediment to action," whereas for
Macbeth it is "the genesis and agency of action."

has pointed out, it should be no surprise that a play about regicide is also about fantasy—for it is in the balance between thinking and doing that the crime of treason hangs.[74]

This link between imagination and political conspiracy might have something to teach us about the pathology of the fantasy as well. Macbeth's murderous hallucination offers a hypothesis about the mechanics of cognition: every act, horrific or otherwise, is imaged in the mind before it is done. The damning hallucinations of *Macbeth*—the dripping dagger and the stubborn spot of blood—represent the mental extrusions of the play's two guilty protagonists. As James Calderwood phrases it, these are "the workings of a self-protective consciousness as it projects inner impulses outward to create a behavioristic world to whose stimuli it can then react."[75] Yet this comes at a cost, Shakespeare shows. Ultimately, the Macbeths become stranded in the mesmeric limbo of intention, the place where past and future are interchangeable—where one may utter the extraordinary statement, "I go, and it is done" (2.1.62). Lady Macbeth is left trapped in this place, always on the brink of murder and also always scrubbing its bloody stain, just as Macbeth finds himself locked within a dull infinity of tomorrows. Being the most heinous of all imaginable offenses, the killing of the king requires the most reckless of ambitions, the staunchest of wills, the most vivid and phantasmagorical of premeditations. Macbeth's hallucinatory motivation therefore draws attention to an aspect of psychopathological discourse glossed over by Shakespeare's medical peers—namely, that imagination, even when it is corrupt—especially if it is corrupt—may be the most startling expression of the mind's agency rather than a symbol of its cowering passivity. It is tempting but also too facile to dismiss the faculty as inhuman, a victim of the humors, slavish to the devil. In this play, the process whereby inward images may be inscribed in outward reality is depicted in the service of an unspeakable crime. But perhaps it is only through the most godless monuments of the imagination that the force of the human will can be affirmed.

The pathologies of *Macbeth* illustrate how deeply imagination was involved in early modern ideas about disorder, whether in regard to the inward balance of bodily humors or the alienating influence of the supernatural. More than that, the play points to the challenges with which Renaissance physicians were wrestling as they sought to refine their etiologies of false perception: the lim-

---

74. Christopher Pye, *The Regal Phantasm: Shakespeare and the Politics of Spectacle* (London: Routledge, 1990), 153–57. See also Kevin Curran, "Feeling Criminal in *Macbeth*," *Criticism* 54, no. 3 (2012): 391–93, which posits that Macbeth's phenomenology blurs the distinction between "criminal thought and criminal acts"; the intent to murder is, for him, equivalent to the act.

75. Calderwood, *If It Were Done*, 38.

its of diagnosis, the construed perimeter around what we take to be an individual, the link between private torment and collective syndrome. There remain the problems of distinguishing the diseases of imagination from its ordinary operations and determining how we relate, or ought to relate, to our unmanageable fancies. It may no longer be possible to speak of the brain as a closed alembic, for to grasp the complexity of a mind that extends beyond the body will require the revision of intuitive assumptions about the psychic interior. It may be thought that the human organism is an enclosed microcosm, needing always to be guarded from the external and internal forces that threaten it. More likely, though, it is the matrix of interlocking effects—both the impressions that infect us from without and those that we devise ourselves and send out into the world—that constitutes the entirety of a person.

CHAPTER 6

Chimeras

Natural History and the Shapes of *The Tempest*

Toward the end of *The Tempest*—a fabulous play of sorcerers, sprites, harpies, and mooncalves—comes a representation of ordinary earthly loveliness. In a marriage masque, the goddesses Ceres, Iris, and Juno pronounce blessings on the newly betrothed Miranda and Ferdinand, wishing the couple "increase" and "plenty."[1] The deities summon images of cultivation and harvest; they sing of "wheat, rye, barley, vetches, oats, and peas," "nibbling sheep," and "flat meads." Speaking of "poll-clipped vineyard[s]," "short-grassed green[s]," and "barns and garners," they invoke rustic toil and seasonal conviviality, human and social warmth; afterward, there is a dance by "sunburned sicklemen" in straw hats (4.1.61–68, 83, 111, 134). However, the truth is that no such agronomic idylls exist on Prospero's island, a place of howling wolves and "ever-angry bears," of "bogs, fens, flats," a place that is careless of human mortality, where a "rich and strange" sea change readily turns bones into coral and eyes into pearls (1.2.289, 2.2.2, 1.2.400). Therefore, the vision that Prospero presents—nature as docile, tidy—is illusory, indeed virtuosic in its illusoriness; it demonstrates, as it is probably meant to, the full extent of this magician's power. It speaks also to the extreme fantastical quality of this play. In the strange world of *The Tempest*, nature and art have

---

1. William Shakespeare, *The Tempest*, 4.1.110. Shakespeare references are from *The Norton Shakespeare*, ed. Stephen Greenblatt et al. (New York: W. W. Norton, 2016), hereafter cited parenthetically.

changed places: here, realistic tableaux of ecological beauty constitute the most dazzling and wondrous artifice imaginable.

Prospero's imagination—he calls the masque an enactment of "my present fancies"—is emblematic of the way this play deploys nature in order to test the dichotomy of imitative and fabulous imagination that prevailed in Renaissance literary theory. With its painterly prettiness, Shakespeare's epithalamic masque is reminiscent of Philip Sidney's idea that the "golden" world of poetry can outdo the "brazen" world of nature: "Nature never set forth the earth in so rich tapestry as divers poets have done; neither with so pleasant rivers, fruitful trees, sweet-smelling flowers, nor whatsoever else may make the too much loved earth more lovely."[2] In the same place, Sidney also says that poetry's relation to nature is unique. Unlike astronomy, music, or rhetoric, which are founded on natural laws and natural forms, poetry is not bounded by nature's "narrow warrant." The poet is able to "grow in effect another nature, in making things either better than nature bringeth forth, or, quite anew, forms such as never were in nature, as the Heroes, Demigods, Cyclops, Chimeras, Furies, and such like." Thus, poetry can either imitate, adorn, and perfect nature, "make it more lovely," or it can bring into being "another nature," fashion "forms such as never were." Shakespeare is here doing both: Prospero's masque, being both a "tapestry" of "sweet-smelling flowers" and an orgy mounted by shape-shifting sprites, inhuman "Chimeras," troubles the distinction between imitation and invention. It does this, specifically, by presenting us with an ambiguous depiction of the natural world, one whose rosy realism is complicated by the implicit interposition of imagination.

Sidney's use of the term *chimera* is worth noting too, for this was a contemporary byword for outlandish phantasms. We have heard it several times already in earlier chapters—used to describe Michel de Montaigne's idle fancies, as well as the specters that bother John Marston's malcontented revenger, and the horrific sights seen by the sinful soul in fancy's mirror. It was commonly understood, in other words, that imagination can spawn monstrous forms—the ideational equivalent of the beast in Greek mythology that is part lion, part goat, and part serpent. The cognitive basis for this is the fantasy's power of conjunction, its ability to stitch together disparate ideas into new wholes. As Sidney's apologia indicates, imagination's power to create "another nature" was not above question and had still to be defended by Renaissance poets. In epistemological terms, mental chimeras are untrustworthy too,

2. Philip Sidney, "The Defence of Poesy," in *The Major Works*, ed. Katherine Duncan-Jones (Oxford: Oxford University Press, 2009), 216.

standing for error and impossibility. In actual reality, the fantasy's combinative force was furthermore a profound physical threat, capable of engendering monstrous animal and human shapes. And yet, as this chapter will suggest, imagination's penchant for chimeric synthesis is everywhere on display in early modern natural history, a protoscientific discourse that was, in the late sixteenth and early seventeenth centuries, weighing the importance of empirical observation and factual representation above emblem and myth. Travel literature and zoological writings of this period employ without knowing it a poetics of description that emulates the chimeric imagination. *The Tempest*, which is more consciously alert to the mediated quality of perception, explores the relation between nature and imagination more obviously. Using a motif of "shape"—Ariel's shape-shifting, Caliban's misshapenness, the shapeliness of Ferdinand and Miranda—Shakespeare's romance exposes the cognitive mechanism by which the mind routinely parses the forms of things, suggesting that if our ordinary reading of nature is always a feat of fancy, then the presumed separation between the fantastic and the icastic may be illusory as well.

Natural-Born Chimeras

The seventh book of Pliny's natural history surveys the "straunge and wondrous shapes of sundrie nations." It tells of the "monsters of men" that may be found at the farthest reaches of the world: one-eyed Cyclopes and cannibalistic Laestrygonians; Troglodites who sleep in the shade of their upturned feet; Hermaphrodites and sorcerers.[3] The chapter detailing these marvels concludes by remarking on the astonishing variegation of the human race. In Philemon Holland's 1601 English translation, the closing lines read thus: "In the deserts of Affricke yee shall meet oftentimes with fairies, appearing in the shape of men and women, but they vanish soone away like fantasticall illusions. See how Nature is disposed for the nones to devise full wittily in this and such like pastimes to play with mankind, thereby not onely to make her selfe merrie, but to set us a wondering at such strange miracles. And I assure you, thus daily and hourly in a manner plaieth she her part, that to recount every one of her sports by themselves, no man is able with all his wit and memorie."[4] About a decade later, Holland's vanishing fairies are echoed in Prospero's "insubstantial pageant" (4.1.155):

---

3. Pliny the Elder, *The historie of the world*, trans. Philemon Holland (London, 1601), 154–55.
4. Pliny, 157.

These our actors,
As I foretold you, were all spirits and
Are melted into air, into thin air. (4.1.148–50)

Holland's lyricism is to his own credit; in Pliny's Latin original, neither "fairies," nor "women," nor "fantasticall illusions" appear.[5] The result is a whimsical characterization of Nature, who is said to "play with mankind" in order to make herself "merrie." Anthropological diversity becomes the creative manifestation of "wit" and "sport," the "daily and hourly" pleasures of an intelligence that transcends the limits of human ingenuity. The text goes on to say that the next chapter will examine further the more peculiar fruits of Nature's recreation; that chapter is entitled "Of prodigious and monstrous births."

Holland's decisions as a translator highlight associations between imagination, natural history, and monstrosity that were active in early modern thinking. The scientific and cultural implications of the contemporary fascination with monsters and marvels have been documented by scholars such as Lorraine Daston and Katharine Park.[6] Doubts and anxieties about imagination are most definitely mixed up in this: perhaps the single most frequently repeated medical detail about the image-making faculty, rivaled only by the stories of melancholy and hallucination we looked at in the last chapter, is the notion that the wandering imaginations of pregnant women can disfigure their unborn children. Pierre Boaistuau is in good company when he notes in his *Histoires prodigieuses* that the "imagination, which the Woman hath, whylest she conceiues the childe," may impress its "beames and Charrecters" on the impressionable form of "the infante."[7] The physician Ambroise Paré's *Des*

5. The original reads *homnium species obviae subinde fiunt momentoque evanescunt*, translated as "ghosts of men suddenly meet the traveller and vanish in a moment" in Pliny, *Natural History, Volume II: Books 3–7*, trans. H. Rackham, Loeb Classical Library 352 (Cambridge, MA: Harvard University Press, 1942), 7.2.32.

6. See Lorraine Daston and Katharine Park, *Wonders and the Order of Nature, 1150–1750* (Cambridge, MA: MIT Press, 2001), chap. 5.

7. Pierre Boaistuau, *Certaine secrete wonders of nature*, trans. Edward Fenton (London, 1569), 96r. Earlier published as *Histoires prodigieuses* (Paris, 1561). Some more examples: Marsilio Ficino, *Platonic Theology*, trans. Michael J. B. Allen, ed. James Hankins (Cambridge, MA: Harvard University Press, 2004), 4:13.1.1: "Look at the various gestures and signs parents impress upon their children . . . and that come from the vehemence of their phantasizing about the various things that happen to affect them when they are mating!"; Girolamo Cardano, *The "De Subtilitate" of Girolamo Cardano*, ed. John M. Forrester (Tempe: Arizona Center for Medieval and Renaissance Studies, 2013), 2:695: "What is quite certain is what many people have frequently suspected, that the condition of pregnant women can damage fetuses in the womb"; Ambroise Paré, *The workes of that famous chirurgion Ambrose Parey*, trans. Thomas Johnson (London, 1634): "That which is strongly conceived in the mind, imprints the force into the infant conceived in the wombe," 978; Levinus Lemnius, *The touchstone of complexions*, trans. Thomas Newton (London, 1576), 93r: "A woman at the time of her conception, stedfastly fixing her ymagination vppon any thinge, deryueth & enduceth certayne markes and tokens thereof

monstres et prodiges, first published in 1573, gathers many examples seen to illustrate this principle. Indeed, Paré presents "imagination" among the key causes of monsters—in among the "glory of God," the "subtlety of the Devill," and circumstances of conception, gestation, and "seed." Not only can the faculty produce localized abnormalities, such as result in conjoined twins or extra limbs, it can radically reinvent the corporeal form. It is easy to see how this phenomenon of maternal impression may be projected onto a feminine allegorization of nature herself. Paré, for instance, takes bizarre marine fauna as a sure sign that Nature "desports it selfe in the framing of them."[8] If, as Helkiah Crooke later writes, a strong imagination "produce[s] formes euen as they say the superiour Intelligences in the Heauens do produce the formes of Mettalles, Plants, and creatures," then perversities of nature may well be put down as the flights of an otherworldly fancy.[9]

There is another, cognitive rather than physical, sense in which anomalous shapes evoke imagination: weird phantasms are themselves akin to monsters, in that they are unusual conglomerations of parts. The ability of the image-making faculty to combine ideas was well known; for Avicenna it had been important enough to warrant splitting the faculty into two powers, one that simply retains information gleaned from the senses and one that can manipulate and rearrange phantasms already formed.[10] This power is described by Thomas Aquinas; paraphrasing Avicenna, he employed a comparison that would be often repeated in early modern writings: "From the image of gold and the image of mountain we compose the single form of a golden mountain which we have never seen."[11] The other much-quoted example is the chimera, the fabulous creature slain by Bellepheron, said to possess the head of a lion, the middle of a goat, and the tail of a serpent—sometimes depicted as having three separate heads (see figure 6). The reason we are able to picture "a Mountain of Gold and the Monster called a *Chimera*, which we have never totally seen," is that "we have seen them by piece-meale."[12]

---

into the Infant"; Pierre de la Primaudaye, *The second part of the French academie*, trans. Thomas Bowes (London, 1594), 157: "The fancies and imaginations of great bellied women are so vehement and violent, that vpon the bodies of the children they goe withall, they print the images and shapes of those things vpon which they haue fixed their fancie."

8. The quotation is from Paré's *Workes*, 1009, book 25 of which gathers the author's earlier writings on monstrosity. The collected works were first published in French in 1575.

9. Helkiah Crooke, *Mikrokosmographia* (London, 1615), 300.

10. Jon McGinnis, *Avicenna* (Oxford: Oxford University Press, 2010), 113–17.

11. Thomas Aquinas, *Summa Theologiæ*, trans. Timothy Suttor, vol. 11 (Cambridge: Cambridge University Press, 2006), 1.78.4. See also André du Laurens, *A discourse of the preseruation of the sight*, trans. Richard Surphlet (London, 1599), 74: "The imagination compoundeth and ioyneth together the formes of things, as of Golde and a mountaine, it maketh a golden mountaine."

12. Daniel Sennert, *Thirteen books of natural philosophy* (London, 1660), 380–81.

FIGURE 6. A chimera rendered with the heads of a lion, goat, and serpent. *The Chimera of Arezzo*, Etruscan bronze sculpture, 400–350 BCE. Photo credit: Scala / Art Resource, NY.

An early use of *chimera* to denote an incredible idea comes in Philippe de Mornay's *Trewnesse of the Christian religion*; it argues that the biblical creation story cannot be an invention of the human "brayne" and that it can only have been handed down "from aboue"—for "how could that Chymera haue come in any mannes mynd?"[13] In the many more instances that might be cited, the chimera trope does not just invoke the relatively neutral notion of ontological impossibility; it particularly evokes monstrosity and grotesqueness. Josuah Sylvester's edition of *The Divine Weeks and Works* glosses "Chymeras" as "strange Fancies, monstrous Imaginations, Castles in the Aire."[14] The underlying logic of the trope is explained by Francis Bacon: in welding together whatever it likes, imagination is free to contravene "the laws of matter," to "join that which nature hath severed, and sever that which nature hath joined, and so make unlawful matches and divorces of things."[15] The

13. Philippe de Mornay, seigneur du Plessis-Marly, *A woorke concerning the trewnesse of the Christian religion*, trans. Philip Sidney and Arthur Golding (London, 1587), 433.

14. Guillaume de Salluste Du Bartas, *Du Bartas his deuine weekes and workes translated*, trans. Josuah Sylvester (London, 1611), Iiir.

15. Francis Bacon, *The Advancement of Learning*, in *The Works of Francis Bacon*, ed. James Spedding, Robert Leslie Ellis, and Douglas Denon Heath (London: Longman, 1859), 3:343.

fruits of this are unnatural, akin to the race of "centaurs and chimeras" birthed by Ixion's coupling with a cloud: "High and vaporous imaginations instead of a laborious and sober inquiry of truth, shall beget hopes and beliefs of strange and impossible shapes."[16] It was evident to Shakespeare's contemporaries that the fantasy can bring forth notional monstrosities as easily as physical ones.

Also contributing to the force of the metaphor, I submit, are the expanding horizons of early modern travel, exploration, and natural history.[17] With information flooding in from all parts of the world, the study of nature greatly widened in scope; scholars strove to incorporate into existing epistemes the flora, fauna, and peoples of, for example, the New World.[18] The middle of the sixteenth century saw the emergence of notable naturalists such as Pierre Belon, Guillaume Rondelet, Joannes Jonstonus, Conrad Gesner, and Leonhart Fuchs. As Brian W. Ogilvie has described, these were "practitioners of a discipline that, though related to medicine and natural philosophy, was distinct from both."[19] Though they followed in the footsteps of Pliny and other authorities, they were keen to correct the mistakes made by the ancients. New methods were becoming the norm: natural history had begun the process, which would culminate later in the seventeenth century, of renovating itself from a mythology of marvels into a science of observation. Plants and animals were examined in their natural surroundings; specimens were systematically collected and cataloged.[20] Increasingly, the new naturalists applied themselves particularly to the matter of representation; description, "both process and result," was their central concern.[21] Images were key in this regard, the idea being that nature, as Fuchs proclaimed, is "fashioned in such a way that everything may be grasped by us in a picture."[22] Also in contrast to their ancient predecessors, early modern natural historians made use of realis-

---

16. Bacon, 3:362.

17. For overviews of natural history in this period, see E. W. Gudger, "The Five Great Naturalists of the Sixteenth Century: Belon, Rondelet, Salviani, Gesner, and Aldovrandi: A Chapter in the History of Ichthyology," *Isis* 22, no. 1 (1934): 21–40; Paula Findlen, "Natural History," in *Cambridge History of Science*, vol. 3, *Early Modern Science*, ed. Katharine Park and Lorraine Daston (Cambridge: Cambridge University Press, 2006), 435–68; and Brian W. Ogilvie, *The Science of Describing: Natural History in Renaissance Europe* (Chicago: University of Chicago Press, 2006), 1–24.

18. J. H. Elliott, *The Old World and the New 1492–1650* (Cambridge: Cambridge University Press, 1970), 39. See as well Marjorie Swann, *Curiosities and Texts: The Culture of Collecting in Early Modern England* (Philadelphia: University of Pennsylvania Press, 2001), 59.

19. Ogilvie, *Science of Describing*, 1.

20. Swann, *Curiosities and Texts*, 57–58.

21. Ogilvie, *Science of Describing*, 6.

22. Leonhart Fuchs, quoted in Sachiko Kusukawa, "Leonhart Fuchs and the Importance of Pictures," *Journal of the History of Ideas* 58, no. 3 (1997): 411. Kusukawa's analysis of the pictures in Fuchs's *Remarkable Commentaries on the History of Plants* shows that the use of images could be as potentially controversial as texts.

tic pictures, drawings done from life rather than stylized depictions, in order to supplement their verbal descriptions.[23] In the transitional moment at which Shakespeare was writing, natural history relied on bibliographical methods as well as firsthand studies; attention to emblematic similitudes sat side by side with an emerging commitment to factual empiricism.

Unconsciously, Renaissance naturalists deploy a descriptive technique, old but effective, that we might call, for want of a better term, chimeric. Here is Pliny on the chameleon: "In shape and quantitie it is made like a Lisard, but that it standeth higher and streighter. . . . The sides, flankes, and bellie, meet togither, as in fishes: it hath likewise sharpe prickles, bearing out upon the backe as they have: snouted it is, for the bignesse not unlike to a swine, with a very long taile thin and pointed at the end, winding round and entangled like to vipers: hooked clawes it hath, and goeth slow, as doth the tortoise: his bodie and skin is rough and scalie, as the crocodiles."[24] The creature is portrayed as a composite of lizard, fish, swine, viper, tortoise, and crocodile. The chameleon is not a monster; yet the many similes working in concert make for a fantastical-seeming whole. Pliny uses this method throughout. The "river horse," or hippopotamus, for example, is said to be taller than a crocodile, with the "cloven foot" of an ox, the "backe, maine, and haire of an horse," and a boar's twining tail.[25] Of the unicorn, or "Monoceros," Pliny writes, "His bodie resembleth an horse, his head a stagge, his feet an Elephant."[26] Something comparable happens with the peoples of Africa and Asia, who are defined in terms of missing or abnormal features. We are told of men who lack noses or mouths, who have "one eye only in the midst of their forehead" or "their feet growing backward."[27] Compare this approach with the very different one taken by Aristotle: the *Parva animalium* is cataloged by bodily organ, not by species; it traverses the animal kingdom according to the many manifestations of beak and horn, heart and lung, epiglottis and diaphragm, to be found therein.[28]

---

23. See William B. Ashworth Jr., "Emblematic Natural History of the Renaissance," in *Cultures of Natural History*, ed. N. Jardine, J. A. Secord, and E. C. Spary (Cambridge: Cambridge University Press, 1996), 24; and Katherine Acheson, "Gesner, Topsell, and the Purposes of Pictures in Early Modern Natural Histories," in *Printed Images in Early Modern Britain: Essays in Interpretation*, ed. Michael Hunter (Farnham, England: Ashgate, 2010), 127–44.

24. Pliny, *Historie of the world*, 215.

25. Pliny, 209–10.

26. Pliny, 206.

27. Pliny, 154.

28. For a sampling, see Aristotle, *Parts of Animals*, trans. A. L. Peck, in *Parts of Animals. Movement of Animals. Progression of Animals*, trans. A. L. Peck and E. S. Forster, Loeb Classical Library 323 (Cambridge, MA: Harvard University Press, 1937), 656a–676a.

Pliny's pattern carries through in subsequent texts. *The Travels of Sir John Mandeville* also iterates through the physical deformations peculiar to different regions: the natives of one island are distinctive for having "no heads" and eyes "in their shoulders"; in another, they have "great and long ears that hang down to their knees."[29] As a sixteenth-century example, take the travelogue of John Leo. Its English translation employs the word *shape* to underline similarities between African and European fauna: apes "represent the shape of man, not onely in their feete and hands, but also in their visages"; the ostrich "in shape resembleth a goose, but that the neck and legges are somewhat longer."[30] In Samuel Purchas, the zebra is said to be an "exquisite" horse, "all ouer-laide with partie coloured Laces, and guards, from head to Taile."[31] Adriaen Coenen, the Dutch beachcomber, writes of a creature that may have been a squid—a fish with two side-fins like the wings of a bat, the curved black beak of a parrot, and eyes as big as a cow's.[32] We see in these texts how imaginative collage helped render otherwise-unfamiliar animal silhouettes for European readers. Ironically, the mechanism of misrepresentation whereby the faculty of imagination was believed to produce unnatural conceits is deployed in natural historical discourse in the service of accurate description.

What happens to the category of the monstrous when what appear to be real-life chimeras are reported to exist? How are we to think of, for instance, the giraffe, a beast with the head of a camel and the specks of a leopard, whose very hybridity is encoded in the portmanteau of its ancient name—"cameleopard"?[33] Paré offers a way to distinguish between monsters and prodigies: the former, he writes, are "what things soever are brought forth contrary to the common decree and order of nature," as, say, a child "borne with one arme alone, or with two heads." Prodigies, on the other hand, are things "altogether differing and dissenting from nature: as, if a woman should bee delivered of a Snake, or a Dogge."[34] For Leo as well, the two groups appear to be distinct: in Africa there are diverse types of eagle, he observes; however, when an eagle couples with a wolf, the outcome is a "monster" with "the

29. John Mandeville, *The Travels of Sir John Mandeville*, ed. E. C. Coleman (Stroud, England: Nonsuch, 2006), 192–93.

30. John Leo, *A geographical historie of Africa*, trans. John Pory (London, 1600), 341, 347.

31. Samuel Purchas, *Purchas his pilgrimage* (London, 1613), 466.

32. Adriaen Coenen, *The Whale Book: Whales and Other Marine Animals as Described by Adriaen Coenen in 1585*, ed. Florike Egmond and Peter Mason (London: Reaktion, 2003), 166.

33. For other citations of this favorite example, see Stephen Batman, *Batman vppon Bartholome his booke "De proprietatibus rerum"* (London, 1582), 353v: "Cameleopardus . . . hath the head of a camell, and the necke of a horse, and legges and feet of a Bull, and specks of the Perde"; and Purchas, *Purchas his pilgrimage*, 464: "The *Giraffa* or *Camelopardis*; a beast . . . of a strange composition, mixed of a Libard, Hart, Buffe, and Camell."

34. Paré, *Workes*, 961.

beake and wings of a birde" and "the feete of a woolfe."[35] Often enough, however, the distinction is hard to see. Paré's chapter on marine monsters contains mermaids, flying fish, and whales; among its evidentiary sources are the reports of travelers such as André Thevet and Jean de Léry, as well as of protoscientists such as Gesner and Rondelet. Like the eclectic wonder cabinets of the period, in which the imaginary, the exotic, and the everyday mingled, Paré's book presents the full spectrum of the natural continuum. In it, the rhetorical device of chimeric description works to unify creation into a marvelous whole.

The device also raises epistemological and classificatory questions. In Walter Raleigh's *History of the world*, for example, apparent similarities between the fauna of America and Europe provoke thoughts about likeness and kinds. Raleigh notes that in the "low *Islands* of *Caribana*" lives a bird resembling the "Crowe and Rooke of *India*" but "full of red feathers." Similarly, there is in Virginia a variant of the "BlackBird and Thrush" whose plumage is "mixt with black and carnation."[36] He weighs the implications of these ornithological findings while considering how Noah's ark could have been large enough to hold all the animals, positing that species diversity may be overly exaggerated: "By discouering of strange landes, wherein there are found diuers beastes and birdes differing in colour or stature from those of these Northerne parts, it may be supposed by a superficiall consideration, that all those which weare red and pyed skinnes, or feathers, are differing from those that are lesse painted, and were plaine russet or blacke: they are much mistaken that so thinke."[37] If the "Ounce of *India*" resembles the "Cat of *Europe*," just as the "Sharke of the South Ocean" looks a bit like the "dogfish of England," this may be because these creatures are not fundamentally distinct. In a noteworthy aside, the English explorer deduces that superficial differences among humans may be equally negligible: "If colour or magnitude made a difference of Species, then were the *Negro's*, which wee call the Blackmores *non animalia rationalia* not men, but some kinde of strange beastes: and so the Giants of the South *America* should be of another kinde, then the people of this part of the World."[38] In a sense, Raleigh is doing as Paré and Leo—trying to decide where to draw the boundary between natural biodiversity and atypical crossbreeding. His inclination here is to presume similarity rather than difference; the chimeric comparisons that suggest themselves to

35. Leo, *Geographical historie*, 348.
36. Walter Raleigh, *The history of the world* (London, 1614), 111.
37. Raleigh, 111.
38. Raleigh, 111–12.

him are, he decides, indicative of overarching biological commonality, well within range of the normal.

Meanwhile, Léry, in his ethnographic account of Brazil, tries to harness the similizing power of *phantasia* in order to communicate with his absent countrymen. To conjure the Tupinamba man, he leads his reader through a series of imaginative exercises: "Imagine in the first place a naked man, well formed and proportioned in his limbs, with all the hair on his body plucked out; his hair shaved in the fashion I have described; the lips and cheeks slit, with pointed bones or green stones set in them; his ears pierced, with pendants in the holes; his body painted; his thighs and legs blackened with the dye that they make from the *genipap* fruit that I mentioned; and with necklaces made up of innumerable little pieces of the big seashell that they call *vignol*. Thus you will see him as he usually is in his country."[39] In the second exercise, the reader is told to denude this savage of "all the flourishes described above" and then "cover his whole torso, arms, and legs with little feathers minced fine, like red-dyed down," thus making him "artificially hairy." In the third instance, attire the man in "garments, headdresses and bracelets." In the fourth, "leave him half-naked and half-dressed," but "give him the breeches and jackets of our colored cloth"; finally, place a maraca and rattles made of fruit in his hand, and see him as "he dances, leaps, and capers about."[40] As Anthony Pagden suggests, there is possibly a quiet despair underwriting this passage, an awareness of the "loss of humanity" that inevitably comes with attempts to make one culture legible to another.[41] Léry wants to share his "autoptic" eyewitness account with those who will never travel to Brazil, but even though his own memories of the Tupi are hauntingly vivid—"I will forever have the idea and image of them in my mind"—it is difficult "to represent them well by writing or by pictures."[42] He intuits the potential of imagination and struggles to actualize it.

Whether we are picturing animals or myths, Africans or Englishmen, the cognitive arithmetic of imagination is involved. For this reason, it is not easy to separate fable and reality, differentiate species, tell an American savage from a European jester. And yet sixteenth-century travelers and natural historians retained their commitment to empirical description and descriptive consistency. The resulting mix of methodological conscientiousness and deferred

39. Jean de Léry, *History of a Voyage to the Land of Brazil*, trans. Janet Whatley (Berkeley: University of California Press, 1990), 62.

40. Léry, 64.

41. Anthony Pagden, *European Encounters with the New World: From Renaissance to Romanticism* (New Haven, CT: Yale University Press, 1993), 51.

42. Léry, *Voyage*, 67.

judgment can be seen in what was perhaps the most significant work of zoography in English. Edward Topsell's *Historie of foure-footed beastes* and *Historie of serpents*, derived from Gesner's *Historiae animalium*, are, like the medieval bestiaries, interested in the cultural histories of animals—their habits, temperaments, relations with one another, and usefulness to humans. As was not unusual, it intersperses mythological beings, even creatures the author strongly suspects do not exist, in among actual ones; in its pages are hounds and horses, as well as the lamia and the Sphinx (see figure 7). Nonetheless, the work clearly aspires to systematic rigor. Each case is considered on the merits of discursive sources and visual evidence. The veracity of the lamia is granted "by the testimony of holy Scripture," whereas the Hydra draws skepticism because the picture is badly drawn: "If it had not been an vnskillful Painters deuice, he might haue framed it in a better fashion, and more credible to the world."[43] However strong or weak the truth claim, the level of detail remains the same: the habitats and feeding habits of dragons warrant as much attention as those of foxes and porcupines. And always there is the procedure of chimeric description, which confers the sheen of the exotic on almost every animal. Fabulous creatures have striking features: the unicorn is identifiable by its horn, the mantichora by its rows of teeth. But the baboon too is remarkable, given that it has a man's body, dog's face, and dragon's beard. The elephant has the color of a mouse, ears like the wings of a bat, and the tail of an ox. The rhinoceros has girdles like the wings of a dragon and skin like the shell of a tortoise.[44] Occasionally the collage effect creates a kind of tautological circularity—the chameleon is like the crocodile, the crocodile is like the chameleon—resulting in a self-referential echo chamber that mutes rather than amplifies the impression of diversity.[45] And the veristic aspirations of Topsell's text do not erase its residual emblematic sensibility, its sensitivity to fable and legend.[46] Its attempt to contain its fantasticality only underlines the obdurate presence of imagination in the history of nature.

Though the chimera evokes the unique style of description we have seen in the zoological catalogs and travel narratives of the late sixteenth and early

---

43. See, respectively, Edward Topsell, *The historie of foure-footed beastes* (London, 1607), 455; and Edward Topsell, *The historie of serpents* (London, 1608), 202.

44. Topsell, *Historie of foure-footed beastes*, 10–11, 192–93, 196, 596.

45. Topsell, *Historie of serpents*, 115, 129. In Shakespeare's *Antony and Cleopatra*, Lepidus asks, "What manner o' thing is your crocodile?" Antony replies, "It is shaped, sir, like itself, and it is as broad as it hath breadth." Lepidus is suitably impressed: "'Tis a strange serpent" (2.7.42–49).

46. William B. Ashworth Jr., "Natural History and the Emblematic World View," in *Reappraisals of the Scientific Revolution*, ed. David C. Lindberg and Robert S. Westman (Cambridge: Cambridge University Press, 1990), 305. Acheson points out that "Topsell was not an expert in natural history, but a clergyman who also published moralising works"; "Gesner, Topsell," 128–29.

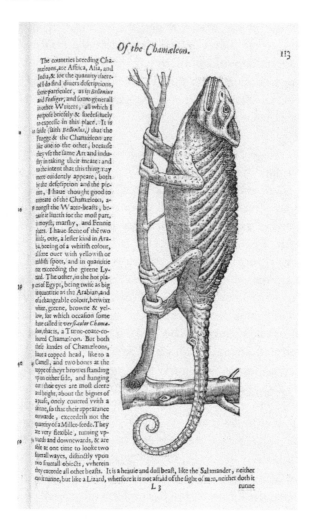

FIGURE 7.   Illustration of chameleon. From Edward Topsell, *The historie of serpents* (London, 1608), 118. RB 17727, Huntington Library, San Marino, California.

seventeenth centuries, the three-headed creature itself is rarely mentioned in these texts. It remains, in a sense, symbolic of the categorically fabulous; it is the definitive, unrepresentable marker of the impossible. With ostensibly real chimeras coming into view, though, the aspiring early modern catalogers of the natural world found in imagination an indispensably useful descriptive instrument. Their unwitting exploitation of the mind's capability for bricolage speaks to the difficulty of overturning mental and rhetorical habits centuries in the making, habits that carry a stubborn, intuitive appeal. What the natural historians do unconsciously, *The Tempest* does explicitly. Exploiting the per-ceptual potentialities of his theatrical art, Shakespeare foregrounds the swift cognitive decisions made in the habitual course of seeing and so lays bare how fundamentally the combinative imagination is involved in our apperception of the shapes of nature.

SHAPES AND THINGS

Imagination has been an enduring topic in the critical tradition of *The Tempest*, not least because of the metaphorical resonances between Prospero's illu-sionistic powers and Shakespeare's dramatic ones.[47] Cognition is, moreover, a theme within the play. For one thing, a seeming fusion of mind and world produces the sense of mutability and potential for transformation that consti-tutes the island's magical atmosphere. The opening tempest is the first of sev-eral instances in which the distinction between organic manifestations and artificial contrivances is blurred. Much of this blurring is the work of Pros-pero, who can translate his "present fancies" into reality, can even "unsettle" the "fancy" of his enemies (4.1.122, 5.1.59); as we saw, the rapturous betrothal masque artfully switches the local wilderness for a bucolic fantasy. Just as in-sistently, though, *The Tempest* suggests the limitations of human intelligence. Prospero is not so wise that he did not once need Caliban to show him the

> qualities o'th' isle:
> The fresh springs, brine-pits, barren place and fertile. (1.2.337–38)

The ecological kingdom of the island is seemingly animated with a panpsy-chic awareness of its own, pulsing with illusions and spirits, with the "sweet airs" of "a thousand twangling instruments" (3.2.129–30). According to fac-ulty psychology, animals possess imagination, which means that, like us, horses, sheep, and goats can dream.[48] Conversely, sleeping humans are temporarily

---

47. As illustrations, see Mary B. Moore, "Wonder, Imagination, and the Matter of Theatre in *The Tempest*," *Philosophy and Literature* 30, no. 2 (2006): 498; Genevieve Guenther, *Magical Imaginations: Instrumental Aesthetics in the English Renaissance* (Toronto: University of Toronto Press, 2012); Andrew Gurr, "The Tempest as Theatrical Magic," in *Revisiting "The Tempest": The Capacity to Signify*, ed. Silvia Bigliazzi and Lisanna Calvi (New York: Palgrave Macmillan, 2014), 33–42.

48. Pliny, *Historie of the world*, 309. See as well Gianfrancesco Pico della Mirandola, *On the Imagi-nation*, trans. and ed. Harry Caplan (New Haven, CT: Yale University Press, 1930), 31: "Nature has endowed the brutes with the capacity to luxuriate in visions and imaginations; of all animate beings man alone reasons, opines, and cognizes."

reconciled with their most creaturely passions and appetites.[49] Still, bestial imagination was not thought capable of forming compound structures such as the mountain of gold; for this, we return to the realm of human cognition.[50] Perhaps because nature and mind are tightly entwined in this play, it has been read as an allegory of the intellect: Prospero, Ariel, and Caliban can be thought of respectively as Platonic reason, will, and appetite, or else as Aristotelian manifestations of the rational, sensitive, and vegetative soul.[51] In short, *The Tempest* is broadly invested in the place of imagination in nature: how it impinges on the lives of people and animals, how it operates within creation.

More specifically, the play points out that our presumptions about the natural world are silently guided by our ideas about the native shapes of the mind. The natural philosopher who seeks to transcribe the laws that undergird nature—disasters, mutations, ecological variegation, the minds of animals— is himself circumscribed by those same laws. Recall that Theseus, in his assessment of imagination, refers to "shape" twice, once to say that lovers and madmen suffer from "shaping fantasies," and once to describe poetry:

> As imagination bodies forth
> The forms of things unknown, the poet's pen
> Turns them to shapes and gives to airy nothing
> A local habitation and a name. (5.1.14–17)

"Shaping fantasies" might be taken to mean, in one sense, that fantasies produce shapes—warp bodies, engender forms. In another sense, fantasies are themselves malleable, taking and abandoning form. Theseus winds up expounding on the second, more phantasmatic sense when he describes the poet's ability to house and christen the "forms of things" through use of poetic language. This remarkable feat Theseus grounds in ordinariness, with ordinary words—a "habitation," a "name"—underlining that the mind's manipulation of "shape" is a quotidian affair, local to human concerns. As we will see, *The Tempest* is interested in both kinds of "shaping fantasies," in both cognitive and corporeal formations, in chimeric figments as well as chimeras, in the usual and the unusual. Relying especially on this notion of "shape," as well as related ideas such as monstrosity, creatureliness, and even thingness,

49. La Primaudaye, *Second part of the French academie*, 158; Du Laurens, *Discourse*, 75; Crooke, *Mikrokosmographia*, 247.

50. Du Laurens, *Discourse*, 75.

51. See, respectively, Murray Wright Bundy, "The Allegory in The Tempest," *Research Studies* 32 (1964): 192–94; and James E. Phillips, "*The Tempest* and the Renaissance Idea of Man," *Shakespeare Quarterly* 15, no. 2 (1964): 147–59.

Shakespeare suggests that while the exercise of imagination can seem strange and produce odd results, it is also a thoroughly normative process by which the "unknown" is molded into intelligibility.

A comic scene in the second act slows and highlights this process as it occurs. Caliban enters with a "gabardine," grumbling about the spirits of the island who constantly pinch and annoy him in the form of "apes," "hedgehogs," and "adders" (2.2.9–13). He sees something approach and, fearing the worst, falls to the ground, hiding under his cloak. Trinculo the jester appears and for a moment puzzles over the shape crouched before him. Then he thinks it may be about to rain. With a blithe philosophical quip—"Misery acquaints a man with strange bedfellows" (2.2.37)—he joins the unidentified creature under the shelter of the gabardine. Now a third person arrives: Stefano, a butler, already tipsy with drink rescued from the shipwreck. He sees a mishmash of limbs and heads and proclaims, "This is some monster of the isle" (2.2.61). Rarely does Shakespearean drama rely so heavily on a prop. The severed hand in *Titus Andronicus*, the forged letter in *Twelfth Night*, even the handkerchief of *Othello*: none of these requires a choreography of actors' bodies quite so precise as this gabardine. *The Tempest* is not short on spectacle; some of its special effects require especially ingenious stage properties. But Caliban's cape is no instrument of wonder; it is a clumsy contrivance, employed for broad comedy. The business with the cloak departs from the sublimity and lyricism that is the play's more usual register. Dramaturgically, it is inessential to the scene, whose purpose is presumably to bring the three would-be conspirators together. And it fails the test of plausibility: Would it not be simpler for Caliban to just flee? Why would Trinculo intrude on the hiding place of a creature he cannot name? Why do they remain enmeshed together for as long as they do?

Only when the scene is read as an ironic theatrical nonillusion that critiques how monstrosity is witnessed do these oddities resolve. This seemingly negligible interlude is no less than a foundational moment in *The Tempest*, not only because in it Shakespeare has fashioned a two-headed, four-limbed monster but also because it exposes how marvels are perceived as such. Marvel seeking is explicitly mentioned by Trinculo as he considers his discovery: "What have we here? A man or a fish? Dead or alive? A fish: he smells like a fish; a very ancient and fishlike smell; a kind of not-of-the-newest poor-john. A strange fish. Were I in England now, as once I was, and had but this fish painted, not a holiday fool there but would give a piece of silver. There would this monster make a man; any strange beast there makes a man. When they will not give a doit to relieve a lame beggar, they will lay out ten to see a dead Indian. Legged like a man, and his fins like arms. Warm, o' my troth! I do now let loose my opinion, hold it no longer: this is no fish, but an islander that hath lately suffered by a thunderbolt" (2.2.24–34). Identification is the initial problem. A number of categories suggest themselves: "man," in contradistinction to "monster" or "beast"; "dead," as opposed to "alive"; "Indian" if not "beggar." After some pondering, Trinculo tries out "strange fish," a classification that suits the olfactory evidence as much as the visual. He then drifts into a profiteering fantasy: "Were I in England now . . . ." Trinculo's intuition that this rare speci-men will be saleable is reasonable; early modern naturalists were known to visit marketplaces to acquire potential objects of study.[52] As though to articu-late the singularity of his find to the prospective "holiday fool," Trinculo mentally disassembles it into a bag of incongruities—animal, human, for-eigner, mutant—that hinder identification and so presumably drive up the market price.

But then Trinculo crawls under Caliban's gabardine. He becomes incorpo-rated into what turns out to be a larger, more elaborate visual joke, architected by the playwright. Caliban, the cowering "islander," is unhappy about his new-found bedmate—"Do not torment me! Oh!" (2.2.53)—but his objections go unnoticed, even though they are spoken in English. The now transfigured "fish" is inspected by another viewer. Stefano's reaction is subtly different from Trinculo's; he too cycles through various classifications, but his have nastier inflections—"savages," "devils" (2.2.54–55). He sees a monster, not a marvel. Whereas Trinculo was hung up on smell, Stefano notices something odd about the creature's voice: "Where the devil should he learn our language? I will give him some relief if it be but for that" (2.2.62–64). Such a voice is worth invest-ing in: Trinculo set his sights on fairground punters, but Stefano thinks he may have found "a present for any emperor" (2.2.65). For the sake of curing the creature's quivering "ague"—a satirizing allusion, perhaps, to the medical knowledge that early modern naturalists often possessed—he proceeds to pour liquor into its gullet. Recognizing Stefano, Trinculo greets his friend, and the spell is at last broken. The "monster" is disarticulated, undergoing a kind of perverse parturition in which the jester is sloppily egested, and Caliban is af-terward branded a "moon-calf"—technically, a stillbirth produced by corrupted seed.[53]

Hereafter, *monster* and *moon-calf* are loudly trilled by Trinculo and Stefano both—ironically so, given that these venal and masterless men are themselves

---

52. William Eamon, "Markets, Piazzas, and Villages," in Park and Daston, *Cambridge History of Science*, 3:210.

53. One contemporary text characterizes a "moone calfe" as "a peece of fleshe without shape growen in the womans wombe, which maketh hir to thinke she is with childe." Thomas Cooper, *Thesaurus linguae Romanae & Britannicae* (London, 1578), Iiii3v.

deviants of a sort. As Mark Thornton Burnett points out, the gabardine monster is suggestive of conjoined twins.[54] But it bears emphasizing that it is not only Caliban under the cloak; Shakespeare deliberately incorporates the boorish travelers in his tableau of monstrosity. The "moon-calf" we see is not Caliban; it is the composite structure made up of all three men, inclusive of the witless posers as well as the gawking viewer. All three are subsequently united in an escalating bestiality, we might note: Caliban is a "puppy-headed" monster; Trinculo is "like a goose," a "jesting monkey," and a "stockfish"; Stefano is a dog (2.2.122; 3.2.17, 42, 67). Their Circean degradation extends to the play's very end, when Ariel lures them like "unbacked colts" through briars and gorse, bringing them "calf-like" before their lords (4.1.176–80).

What has been set on stage for us to see is not a monster but rather the activity of looking at monsters. Through the interpretive predispositions of these clowns, and using the inherently perspectival quality of stage drama, Shakespeare reveals the combinative imagination at work, and shows the cognitive assembly of shapes whereby the mind casually makes and unmakes chimeras. The gabardine mix-up is the first of this play's illusions; yet to come are many more tricks that will awe the audience and onstage onlookers alike. This initial trick, however, reveals itself—reveals, in a way, how all the other tricks will be played; it is a daring gambit, in which the playwright-magician uncovers his device. And, more than just a skilled dramatist's flourish, this is an uncanny scene. For here is the garbled optics of encounter, a teasing and maybe troubling reenactment of what possibly lurks behind those earliest premodern travelogues—the hazy mirages, witnessed by those long-ago explorers who, vexed by privation and alarm, came to believe that they had met with ogres and fairies at the far sides of the world.

Caliban's physical shape has attracted a great deal of critical interest in its own right. The invitation to speculate is writ large across the text, which offers up a bewildering array of epithets: "freckled whelp," "earth," "tortoise," "slave," "plain fish," "demi-devil" (1.2.283, 314, 316, 319; 5.1.269, 275). It can be no accident that none of these is uttered by Caliban himself, who never alludes to his physique; Shakespeare may be suggesting that the matter of Caliban's biology is more our problem than his. The unknowability of Caliban's essential nature in part formulates a question about what counts as human, a designation that *The Tempest* interrogates with nuance by way of thought-provoking approximations—the superhuman protagonist, his gentle and humane daughter, their unhuman attendant sprite, their conniving, inhuman

---

54. Mark Thornton Burnett, *Constructing "Monsters" in Shakespearean Drama and Early Modern Culture* (Basingstoke, England: Palgrave Macmillan, 2002), 136.

political enemies.[55] Even more generally, we might ask where Caliban is sup-posed to fit "on the scale of being."[56] If we cannot decide among "man, sav-age, ape, water-beast, dragon," or "semi-devil," then it may mean, as G. Wilson Knight says, that "Caliban is all of them."[57] He in some sense embodies the very idea of nature, nature as opposed to "civilized order," nature as distinct from "nurture," nature as pure "creatureliness."[58]

Taking into account the careful orchestration of the gabardine illusion, however, it is evident that Shakespeare is as interested in Caliban's silhouette as in his essential or even symbolic nature, interested in how we, quite literally, see Caliban. We know that Richard III is "bunch-backed," and we can easily imagine Bottom wearing an ass's nole. But what does Caliban look like? The many possibilities provided in the text pose an interesting challenge in produc-tion. How is an actor to figure this chimerical medley and become part tor-toise, part fish, part demon? He must choose among the different Calibans on offer, make a new one up, or else strive for a sort of noncommittal ambigui-ty.[59] The promise of a director's singular perspective might seem to resolve the problem, but really it only exacerbates the tension, for whatever interpre-tation the players have chosen for us, we as spectators bear the cognitive bur-den of aligning it with the words Shakespeare has written. However we have decided to solve the riddle, the riddle is asked of us again and again, with each new aquatic or reptilian slur that is thrown at Caliban. Subtly, persistently, and pointedly, the spectacle of Caliban in performance nags at the very mental fac-ulty believed in Shakespeare's time to be responsible for editing phantasms.

Underlining this fluidity is the play's frequent use of the word *shape*. Cali-ban, it is said, is "not honored with / A human shape" (1.2.283–84). He is a

55. In support of Caliban's humanity, see Alden T. Vaughan and Virginia Mason Vaughan, *Shake-speare's Caliban: A Cultural History* (Cambridge: Cambridge University Press, 1991), 10–15. See also Bruce Boehrer, *Shakespeare among the Animals: Nature and Society in the Drama of Early Modern England* (New York: Palgrave, 2002), 30; and Andreas Höfele, *Stage, Stake, and Scaffold* (Oxford: Oxford Univer-sity Press, 2011), 242–43.

56. Barbara Howard Traister, *Heavenly Necromancers: The Magician in English Renaissance Drama* (Columbia: University of Missouri Press, 1984), 140.

57. G. Wilson Knight, *The Crown of Life: Essays in Interpretation of Shakespeare's Final Plays* (Ox-ford: Oxford University Press, 1947), 212.

58. See, respectively, Derek Traversi, *The Literary Imagination: Studies in Dante, Chaucer, and Shake-speare* (East Brunswick, NJ: Associated University Presses, 1982), 233; Northrop Frye, *Northrop Frye's Writings on Shakespeare and the Renaissance*, ed. Troni Y. Grande and Garry Sherbert (Toronto: Univer-sity of Toronto Press, 2010), 46; and Julia Reinhard Lupton, "Creature Caliban," *Shakespeare Quarterly* 51, no. 1 (2000): 8.

59. For a survey of decisions made by actors and directors in the past, see Virginia Mason Vaughan, "'Something Rich and Strange': Caliban's Theatrical Metamorphoses," *Shakespeare Quar-terly* 36, no. 4 (1985): 390–405.

"misshapen knave" (5.1.271), "as disproportioned in his manners / As in his shape" (5.1.292–93). There are other sorts of shapes on the island. When spirits arrive bearing a fantastic banquet, Gonzalo remarks on their "monstrous shape" (3.3.32), echoing the stage direction, which also refers to "strange shapes." Later, Caliban, Trinculo, and Stefano are pursued by Ariel's goblins "in shape of dogs and hounds" (4.1.251 s.d.). Ariel's own shape is as amorphous as Caliban's; his tragedy is not that he is saddled with deformity but that he is too easily able to alter himself, Prospero periodically demanding that he take the "shape" of a harpy or become a "shape invisible" (4.1.185). The word *thing* operates similarly: Ariel is a "Dull thing" (1.2.285), Caliban a "thing most brutish" (1.2.356). When Trinculo and Stefano are reunited with their masters at the end of the play, Sebastian smirks, "What things are these?" Echoing Trinculo's earlier response to Caliban, Antonio retorts, "One of them / Is a plain fish, and no doubt marketable" (5.1.268–69). Consistently, *shape* and *thing* index the mutability not just of physical forms but also of our perception of those forms. Compare the woolliness of these terms with the relative solidity of *monster*. The fifty or so instances of *monster* are all uttered by the marooned Europeans, who use it to signal the unexpected or unfamiliar: "'twas a din to fright a monster's ear" (2.1.307); "Oh, it is monstrous, monstrous!" (3.3.96). Unhappy and disoriented, they have monsters on the mind: Sebastian speaks of unicorns and the phoenix (3.3.23–24); Gonzalo talks of "mountaineers, / Dewlapped like bulls" and "men / Whose heads stood in their breasts" (3.3.45–48). They conjure the language of monstrosity unthinkingly, as a way of regaining "psychological composure."[60]

But Shakespeare's deliberately fuzzy characterization of Caliban seems surely to call for a more conscious, more cautious deployment of imagination than this. Whereas *monster* forecloses interpretation, *shape* appears to invite or defer it. The difference is between accident and evidence, between hearing a "warning"—the Latin root of *monster*—and, more neutrally, encountering a "form"—the Old English source of *shape*. *The Tempest* advocates a kind of perceptual suspension whereby the comprehension of forms is put off long enough so that we may glimpse the mental magic that resolves shapes and things into recognizable ideas. There are applications of this not just for the science of describing but also the description of aesthetic beauty. That shape reading in itself can be an act of representation was understood by Leonardo da Vinci, for example, who knew that in clouds, stains on walls, and other unlikely places, the artist's imagination finds pictures of marvelous landscapes,

---

60. Burnett, *Constructing "Monsters,"* 2.

demons, and dragons.[61] The chimeric interpretation of shapes, therefore, has as much to do with effective art as good science.

BEAUTEOUS CREATURES

Plato's distinction between the icastic and fantastic imagination, which privileges the making of likenesses above the fabrication of appearances and had had among its proponents Marsilio Ficino and Torquato Tasso, continued to exert considerable influence on sixteenth-century aesthetics.[62] George Puttenham, for example, emphasizes the certain division between the beauteous and the monstrous: "The fantastical part of man (if it be not disordered) [is] a representer of the best, most comely, and beautiful images or appearances of things to the soul and according to the very truth. If otherwise, then doth it breed chimeras and monsters in man's imaginations, and not only in his imaginations, but also in all his ordinary actions and life which ensues."[63] Puttenham uses "chimeras" to refer to fantastical notions next door to "monsters." Yet, as we have seen, the chimera encapsulates a poetics of perception that, however fantastical, was mobilized within cultures of natural history not to breed monsters but rather to detail living forms with efficiency and precision. The chimeric poetics of Renaissance travel writing and zoography elide the categories that Puttenham assumes.

If naturalness was to be prized in Renaissance poetry, then this was conceivably complicated by the fact that it was not evident just what counted as natural, as opposed to artificial. Daston and Park write that the *Wunderkammern* of the early seventeenth century "helped transform the ontological categories of art and nature" as they were understood in natural history and natural philosophy. Monstrosity could be considered as art, nature herself a virtuoso "artisan."[64] Imagination seems to have been instrumental in this cat-

---

61. Leonardo da Vinci, "Treatise on Painting," in *Leonardo on Painting*, trans. Martin Kemp and Margaret Walker, ed. Martin Kemp (New Haven, CT: Yale University Press, 1989), 201, 222. Leonardo's idea is paraphrased in Giovanni Battista Armenini, *On the True Precepts of the Art of Painting*, ed. Edward J. Olszewski (New York: Burt Franklin, 1977), 262: "When one gazes at these spots, diverse fantasies and strange fanciful forms seem to take shape." On the history of this notion, see H. W. Janson, "The 'Image Made by Chance' in Renaissance Thought," in *De Artibus Opuscula XL: Essays in Honor of Erwin Panofsky*, ed. Millard Meiss (New York: New York University Press, 1961), 256–58.

62. See J. M. Cocking, *Imagination: A Study in the History of Ideas*, ed. Penelope Murray (London: Routledge, 1991), 206; Torquato Tasso, *Discourses on the Heroic Poem*, trans. Mariella Cavalchini and Irene Samuel (Oxford: Clarendon, 1973), 29.

63. George Puttenham, *The Art of English Poesy*, ed. Frank Whigham and Wayne A. Rebhorn (Ithaca, NY: Cornell University Press, 2007), 110.

64. Daston and Park, *Wonders and the Order*, 209, 261. See too Michael Witmore, *Culture of Accidents: Unexpected Knowledges in Early Modern England* (Stanford: Stanford University Press, 2002), 121,

egorical confusion. On the one hand, nature is fabulous, a fantasist dreaming forth mermaids and desert-dwelling fairies. On the other hand, maternal impression is, just as counterintuitively, a kind of monstrous imitation—wherein the mother becomes, as Marie-Hélène Huet nicely puts it, an "artist of blind resemblances," one who copies without "choice, adaptation, or discrimination" and in doing so produces not a beautiful but a monstrous result.[65] Perhaps it is no surprise that the aesthetic value of the fantastical imagination was being revised during this period, with Neoplatonists and literary critics increasingly disposed to prize invention above "the slavish imitation of nature."[66] An instance of this is the one with which we began: Sidney's apology for poetry, while generally privileging poetry that is *eikastiké*—"figuring forth good things"—also praises the poet's unique ability to bring forth "forms such as never were in nature."[67]

Artistic practitioners traversed such distinctions without compunction. Leonardo, for instance, understood that the fabulous could be portrayed realistically just as, conversely, nature can be made to appear fantastical. Giorgio Vasari tells of how the artist once fashioned a living dragon out of a lizard: "He fastened scales taken from other lizards, dipped in quicksilver, which trembled as it moved, and after giving it eyes, a horn and a beard, he tamed it and kept it in a box."[68] Leonardo also held that mythical beings could be drawn by selectively copying the features of ordinary animals: "You know that you cannot invent animals without limbs, each of which, in itself, must resemble those of some other animal. Hence if you wish to make an animal, imagined by you, appear natural—let us say a Dragon—take for its head that of a mastiff or hound, with the eyes of a cat, the ears of a porcupine, the nose of a greyhound, the brow of a lion, the temples of an old cock, the neck of [a] water tortoise."[69] This instruction, which proposes the chimerical method we have already seen in Pliny and Topsell, issues a reminder that the mental apparatus of both the artist and the scientist are one and the same. *Phantasia* for Leonardo was "the ability to recombine images or parts of images into wholly

---

which suggests that "nature's artifice in singularities" was also tied to the notion of experiment, clarifying for Francis Bacon the role of chance in induction.

65. Marie-Hélène Huet, *Monstrous Imagination* (Cambridge, MA: Harvard University Press, 1993), 26.

66. Daston and Park, *Wonders and the Order*, 210. See also Martin Kemp, "From 'Mimesis' to 'Fantasia': The Quattrocento Vocabulary of Creation, Inspiration and Genius in the Visual Arts," *Viator* 8 (1977): 347–98.

67. Sidney, "Defence of Poesy," 216, 236.

68. Giorgio Vasari, *The Lives of Painters, Sculptors and Architects*, ed. William Gaunt (London: Dent, 1963), 2:166.

69. Leonardo da Vinci, *Leonardo's Notebooks: Writing and Art of the Great Master*, ed. H. Anna Suh (New York: Black Dog and Leventhal, 2013), 166.

new compounds or ideas."[70] Yet his fantastical inventions, Martin Kemp notes, are never inharmonious or implausible, "in their composition deriving from the causes and effects of the natural world."[71] Thomas DaCosta Kaufmann believes that Leonardo may have inspired the extraordinary composite portraits of Giuseppe Arcimboldo, in which many living creatures are pressed together in the form of a human subject (see figure 8).[72] Like optical illusions, the paintings draw the viewer's eye alternately to the parts and the whole. Renaissance art theorists praised Arcimboldo for the clever way he straddles the real and unreal, the beautiful and ugly. Giovan Paolo Lomazzo deemed the rendering of Emperor Maximilian II's vice-chancellor "a marvel" in itself.[73] Gregorio Comanini observed that even though the component parts—foods, beasts, books—are drawn naturalistically, the overall effect is undeniably "fantastic." Comanini and others found these pieces to be grotesque, dreamlike, and chimerical.[74]

Shakespeare toys with the distinction between the chimeric and the aesthetic in Ferdinand's wooing of Miranda. Their courtship is presented as a mutual instruction, a learning to perceive, with each struggling to interpret the other's appearance as a composite and collection of parts. In keeping with the diction of the play, their compliments to one another employ nebulous terms. Ferdinand calls Miranda a "precious creature" (3.1.25); she thinks him

a thing divine, for nothing natural
I ever saw so noble. (1.2.417–18)

To Ferdinand's mind, Miranda's beauty surpasses all others' because it encompasses theirs:

For several virtues
Have I liked several women; never any
With so full soul but some defect in her

70. Carmen C. Bambach, ed., *Leonardo da Vinci: Master Draftsman* (New Haven, CT: Yale University Press, 2003), 26.

71. Martin Kemp, *Leonardo da Vinci: The Marvellous Works of Nature and Man* (Cambridge, MA: Harvard University Press, 1981), 161.

72. Thomas DaCosta Kaufmann, *Arcimboldo: Visual Jokes, Natural History, and Still-Life Painting* (Chicago: University of Chicago Press, 2009), 34.

73. Giovan Paolo Lomazzo, *Idea of the Temple of Painting*, trans. and ed. Jean Julia Chai (University Park: Pennsylvania State University Press, 2013), 164.

74. See Thomas DaCosta Kaufmann's discussion of Comanini in *The Mastery of Nature: Aspects of Art, Science, and Humanism in the Renaissance* (Princeton, NJ: Princeton University Press, 1993), 102, 107, 163.

FIGURE 8. Giuseppe Arcimboldo, *Water*, 1566. Photo credit: Erich Lessing / Art Resource, NY.

> Did quarrel with the noblest grace she owed
> And put it to the foil. But you, O you,
> So perfect and so peerless, are created
> Of every creature's best. (3.1.42–48)

Miranda is a piecewise paragon, possessing a perfection made up of "every creature's best." It sounds a little like emblazoned beauty, the female body disarticulated in a manner Shakespeare decisively lampoons in sonnet 130: "My mistress' eyes are nothing like the sun." But none of Miranda's parts are explicitly described, and the word "creature" holds the indelicacy of comparison slightly at bay, rescuing the speech from cliché. Here is Miranda's reply:

> I do not know
> One of my sex; no woman's face remember
> Save, from my glass, mine own. Nor have I seen

More that I may call men than you, good friend,
And my dear father. How features are abroad
I am skilless of; but by my modesty,
The jewel in my dower, I would not wish
Any companion in the world but you,
Nor can imagination form a shape
Besides yourself to like of. (3.1.48–57)

Miranda's problem, we realize, is the opposite one: she has no experience on which to base an assessment of Ferdinand's looks. Earlier, she was gently mocked for her manifestly awed response to Ferdinand's physique: "Thou think'st there is no more such shapes as he," joked Prospero, "having seen but him and Caliban" (1.2.477–78).

In their separate ways—Miranda through years of exilic isolation and Ferdinand by virtue of being the seemingly sole survivor of a shipwreck—these lovers have been drawn away from the traditional world of Aristotelian universals, says Elizabeth Spiller. As they subsequently strive to wrest meaning from singularities and particulars, they enact the early modern period's paradigmatic shift from "Aquinian wonder to a Baconian knowledge."[75] They are scientists, not philosophers, in that they take prodigies of nature—that is to say, each other—as evidentiary facts rather than aberrant signs.[76] How differently, though, they proceed as explorers; how complementary their registers of perception. Like Pliny and Paré, Ferdinand disassembles his subject in order to render her excellence describable. We might notice that, in doing this, he reinscribes such categories as "defect" and "best," "grace" and "foil." Miranda is a different sort of ethnographer. She proceeds with illimitable caution, and her deepening admiration of Ferdinand is tempered by a simultaneous reevaluation of her own appearance. She grasps that there are men "abroad," though she cannot guess at their "features." She knows she cannot speak of other women and so speaks only of herself. She knows she has little to say about beauty and so speaks instead of intelligence, understanding that the "skill" by which Ferdinand perceives her is something that he has learned and she not yet. She invents a sublime aesthetic out of her ignorance, instinctively choosing to love that which exceeds the "shape" of "imagination." In this phantasmatic exchange between Miranda and Ferdinand, haunted by all the

---

75. Elizabeth Spiller, "Shakespeare and the Making of Early Modern Science: Resituating Prospero's Art," *South Central Review* 26, no. 1 (2009): 32.

76. Lorraine Daston, "Marvelous Facts and Miraculous Evidence in Early Modern Europe," in *Wonders, Marvels, and Monsters in Early Modern Culture*, ed. Peter G. Platt (Newark: University of Delaware Press, 1999), 77, describes the difference.

faces that he has seen and all the ones that she has not, we witness how the chimeric imagination may be used to mediate vision deftly rather than crudely, as we gaze on the unfamiliar.

Fittingly, the play's final scene reinforces the importance of this mediation. Miranda, for one, is confronted with a great many new faces:

> Oh, wonder!
> How many goodly creatures are there here!
> How beauteous mankind is! Oh, brave new world
> That has such people in't! (5.1.181–84)

The gentle irony of this proclamation, ironic because the greater part of the play has presented the "goodly creatures" behaving somewhat badly, evaporates under the warmth of Miranda's admiration, which characteristically valorizes what she sees using the ennobling marker of the creaturely. Before our eyes, her wonder generously reshapes these weary, bedraggled Italians into a celebration of "mankind." Perhaps because of this generosity, she herself is mistaken for a deity: "Is she the goddess that hath severed us," asks Alonso, "and brought us thus together?" (5.1.188–89). When Caliban is brought in, he, very much like Miranda, is struck by the diverse beauty of the scene, moved to think of his god:

> O Setebos, these be brave spirits indeed!
> How fine my master is! (5.1.264–65)

The "fine" master, however, now steps forward to present an uncharitable biography of the slave; Prospero puts Caliban through a final set of phantasmatic contortions, calling him the son of a "witch," a contorted "knave," "bastard," and "demi-devil" (5.1.272–76). And yet Prospero also professes to both "know and own" Caliban:

> This thing of darkness I
> Acknowledge mine. (5.1.278–79)

With the "I" dangling at the end of the line, the "darkness" of Caliban belongs for a small moment to Prospero; they are, just for an instant, conjoined in their thingness. This final scene of recognition, then, is filled as well with misrecognitions; gods, kings, fools, and spirits are mistaken for one another. If the world seems wondrous to us, it is partly so because of the ease and rapidity with which imagination renders such transformations.

If mental representation is fundamentally chimeric, then to see newly may mean to see slowly, or multiply, or indistinctly—to see shapes and things instead of mooncalves and monsters. In turn, this implies that the usual distinction between imagination's two faces, between the faithful and the freakish, may be in need of reconsideration. Of course, it cannot be said that at the end of *The Tempest* this distinction is erased: Caliban remains the misshapen knave, after all, and Miranda the comely creature. The faculty of imagination, likewise, would retain for some time more its associations of distortion and error. As the natural historical discourse of the period indicates, imperatives to observe the world in new terms do not demolish old habits of seeing and thinking shaped by venerable tropes and classificatory traditions. Nonetheless, Shakespeare has in Prospero's island grown a fabulous world, "another nature," that paradoxically illuminates how we see ordinarily in our own. To think about the shape of nature, we learn, means to confront the structure of human psychology as we have conceived it. In *The Tempest*, imagination is not so much a systematic failing as a predictable cognitive procedure; the challenge lies in acknowledging our perceptual creatures as our own.

Epilogue
The Rude Fantasticals

In the preceding chapters I have sought to show how we might read Shakespeare with the aid of early modern imaginative theory and early modern epistemes, so as to recognize that his art was enriched as much by problems of psychology as by aesthetic customs or ethical precepts. While Shakespeare's treatment of imagination is certainly to be counted as part of the general Renaissance reassessment of fancy's merits, it also moves past the assumption that phantasms are either not of this world or else too much of the world; the ultimate inclination is not to mystify but to investigate mental representation, not to judge but to portray its nature. Writing in a consequential period of protoscientific reorientation, Shakespeare appears to have sensed the cultural tremors that were produced, intuiting that in the uncertain intellectual fortunes of the imagination lay an opportunity to expand its aesthetic and literary relevance. Throughout his works, he showcases the unique interceding power of the image-making faculty, whose porous discourse and pliable conception could be used to negotiate—to clarify, illuminate, rethink—novel ambiguities and transitions arising in the conventional understanding of nature, knowledge, and humankind. Shakespeare's plays and poems are important not only in the history of imagination but also to the history of representation, for they decisively render the complicated quality of psychology a kind of beauty. There are more things in heaven and earth than are dreamt of in philosophy, Hamlet says, but this does not dissuade him or anyone else from philosophizing. No small part of the drama of experience, show Shakespeare's Miranda and Venus and Biron, unfolds in the difficult perceptual calculations we are regularly required to make, make with ingenuity or exuberance or dread. As the phenomenon and also the idea that occasions these calculations, imagination is a productive, indeed artistic force.

Why, then, has Shakespeare's imagination not chiefly been associated with knowledge making? Speculating on this question, in this epilogue I will suggest that the answer has something to do with the theater. It seems safe to say that Shakespeare's preferred medium was highly instrumental to his intuitions about imagination. Many of his most phantasmatic episodes are metadramatic, self-reflexively calling attention to the perceptual quality of stage drama—we see as much with the spectacle of Lady Macbeth's sleepwalking or in Edgar's conjuration of a Dover seascape. I want to suggest that the possibilities of the dramatic form crucially melded with faculty psychology and early science in a unique confluence that aided Shakespeare toward his particular perspective of imagination; this confluence might in part explain why Shakespeare's epistemological interest in the faculty was obscured—or at least differently conceived—in the two centuries after his death. To see this, we will first return to *A Midsummer Night's Dream* and explore the mechanical imagination that Shakespeare there foregrounds in the play-within-play staged by Bottom and the artisans. We will then turn to a consideration of how the mechanics of imagination were sequestered after the Renaissance, when both the idea of imagination and Shakespeare's literary identity were strategically remade.

It is not Theseus and Hippolyta but the "rude mechanicals" who have the final word on imagination in Shakespeare's fairy comedy.[1] In a superficial sense, the artisans' clunking rendition of "Pyramus and Thisbe" is extraneous to the play. However, if, as I have been urging, we consider imagination in nonaesthetic contexts, then the mechanicals' theatrical turn emerges as yet another sign of Shakespeare's interest in imagination as a cognitive device. From a conventional perspective, Quince's men seem not to understand imagination at all. Yet if we take them not as unimaginative thespians but instead as imaginative craftsmen—"rude fantasticals," so to speak—then their performance becomes an oblique continuation of Theseus and Hippolyta's earlier conversation about "fancy's images," with which this book began. These artisan-actors differ in a crucial respect from Shakespeare's other imaginers—unlike Juliet or Lear, their imaginations are mobilized in the ser-

---

1. William Shakespeare, *A Midsummer Night's Dream*, 3.2.9. Shakespeare references are from *The Norton Shakespeare*, ed. Stephen Greenblatt et al. (New York: W. W. Norton, 2016), hereafter cited parenthetically.

vice of playmaking. Because of this, they provide insight into Shakespeare's thinking about the relation between imagination and theater.

Shakespeare underlines the fact that the actors charged with delivering Theseus's nuptial entertainment are amateurs—humble and "hard-handed men" who have never "labored in their minds" (5.1.72–73). Driving the point, he inscribes their day jobs in their names—the carpenter's "quince," the weaver's "bottom," the bellows' whistling "flute." As Patricia Parker has described, Puck's descriptor for these manual workers, "rude mechanicals," instantly situates them within certain social and intellectual frames, aligning them with the artifactual over the natural, the lowly rather than the gentle, and the practical rather than the contemplative.[2] To these various associations we could add imagination—or, rather, a lack of imagination. Historically, the mechanical labor of craftspeople was likened to the work performed by animals and slaves: Aristotle, for instance, writes that "artisans" do things unthinkingly, "without knowing what they are doing."[3] In medieval disciplinary classifications, the mechanical arts—which include fabric making, armament, commerce, agriculture, hunting, medicine, and sometimes theatrics—were purposely distinguished from the more noble liberal arts.[4]

As with other areas of knowledge making, the *artes mechanicae* saw advances in the sixteenth century.[5] The previously scattered work of ancient authorities was consolidated in new translations and commentaries.[6] Renaissance engineers made innovations in many fields—Leon Battista Alberti in architecture, Albrecht Dürer in printmaking, and Niccolò Fontana Tartaglia in ballistics, to name only a few. The mathematician Guidobaldo del Monte was among those who moved to rehabilitate the prestige of mechanics. Stressing that simple machines such as the balance, pulley, and wheel could rival the laws of nature, his *Mechanicorum liber* urges that the terms *mechanic* and *engineer* be purged of their pejorative connotations. In so

---

2. Patricia Parker, "Rude Mechanicals," in *Subject and Object in Renaissance Culture*, ed. Margreta de Grazia, Maureen Quilligan, and Peter Stallybrass (Cambridge: Cambridge University Press, 1996), 45.

3. Aristotle, *Metaphysics, Volume I: Books 1–9*, trans. Hugh Tredennick, Loeb Classical Library 271 (Cambridge, MA: Harvard University Press, 1933), 981b.

4. Hugh of Saint Victor, *The Didascalicon of Hugh of Saint Victor*, trans. Jerome Taylor (New York: Columbia University Press, 1991), 54. For more on the classical and medieval conception of the mechanical arts, see George Ovitt Jr., *The Restoration of Perfection: Labor and Technology in Medieval Culture* (New Brunswick, NJ: Rutgers University Press, 1987), esp. chap. 4.

5. Elspeth Whitney argues that the recuperation of the mechanical arts began in the Middle Ages, in fact; see her *Paradise Restored: The Mechanical Arts from Antiquity through the Thirteenth Century* (Philadelphia: American Philosophical Society, 1990).

6. For example, Frederico Commandino produced an edition of Archimedes; Niccolò Fontana Tartaglia, Francisco Toledo, and Giambattista Benedetti wrote commentaries on Aristotle's *Physics*. See A. Rupert Hall, *The Revolution in Science, 1500–1750* (London: Routledge, 1983), chap. 4.

arguing, however, Guidobaldo preserves the distinction between elite engineers and inexpert laborers. Though he concedes that mechanics necessarily
relies on the expertise of the builder and carpenter and mason, he stresses
that mathematicians experience machines in a fundamentally different way
from workmen. "Manual workers, builders, carriers, farmers, sailors, and
many others" are characterized as the beneficiaries, not the producers, of
mechanical science.[7]

Predictably, the notion of imagination straddles the border between the
lofty and lowly aspects of mechanical work. Pierre de la Primaudaye describes
fancy as a mental tinker: it "cutteth asunder and seweth vp againe"; it can
"counterfaite," "worke vpon," and "build"; it "forgeth and coyneth." Like any
artisan, imagination cannot truly create; it can only imitate the "paterne" of
nature and "the woorkes of God."[8] In other accounts, imagination is a special ability that separates the designer from the maker. John Dee's paraphrase
of Alberti emphasizes that the exemplary architect is talented in fantasy and
is able to assemble his creation in "minde and Imagination." He that performs the physical construction is only a "Carpenter"—the "Architectes
Instrument."[9] In the context of artistic production, imagination is necessary to
the prolonged process of mental draftsmanship that comes after the fabled
moment of inception. Franciscus Junius says in his *Painting of the ancients* that
good artificers imagine the making of a work "as if they were present at the
doing, or saw it acted before their eyes"; they immerse themselves in "quick
and lively imagination" until it is time to disburden "their overcharged
braines by a speedie pourtraying of the conceit."[10] Thus, whether imagination is aligned with applied mathematics or unmathematical skill, with intellective rigor or manual craft, it calls attention to the divide between science
and artisanry—the slender separation that early modern mechanics was trying hard to articulate.[11] *Mechanical* is almost a term that implies its own contradiction, denoting both the presence and absence of real knowledge, as well
as of imagination.

---

7. See the prefaces by Guidobaldo and Filippo Pigafetta in *Mechanicorum liber*, in *Mechanics in Sixteenth-Century Italy*, trans. Stillman Drake and I. E. Drabkin (Madison: University of Wisconsin Press, 1969), 241; see as well 248, 256. Originally published in 1577. On Guidobaldo's attitude toward workmen, see M. Henninger-Voss, "Working Machines and Noble Mechanics: Guidobaldo del Monte and the Translation of Knowledge," *Isis* 91, no. 2 (2000): 233–59.

8. Pierre de la Primaudaye, *The second part of the French academie*, trans. Thomas Bowes (London, 1594), 155–56.

9. This is John Dee's English paraphrasing; see his preface to Euclid, *The Elements of geometrie*, trans. Henry Billingsley (London, 1570), d4r.

10. Franciscus Junius, *The painting of the ancients* (London, 1638), 62.

11. Dee, preface to Euclid, *Elements*, a3v.

By taking to the stage, Shakespeare's mechanicals do not just bravely re-tool in a new *techne*; they mobilize that faculty of predictive design that work-men were presumed not to possess. We can see this because Shakespeare shows us "Pyramus and Thisbe" in rehearsal as well as performance—something he does not do with the pageant of the Nine Worthies in *Love's Labor's Lost*, say, or Hamlet's *Mousetrap*. In the rehearsal scenes, the artisans—they call one another "masters"—proceed in an orderly and focused way (1.2.13, 81; 4.2.15). Attendance is taken ("answer as I call you"), materials distributed ("here are your parts"), key tasks allocated ("con them by tomorrow night"), and the next meeting set ("in the palace wood") (1.2.14, 81–84). To be sure, there are some problems: Snug is slow of study, Flute has trouble following the script, and the male lead is spirited away by fairies because he "forsook his scene" (3.2.15). Yet when it comes to imaginatively visualizing the production, the men are adept. They anticipate the problems that may arise as though "present at the doing"—and dispatch these one by one. The onstage killing, which "the la-dies cannot abide," may be resolved by "leav[ing] the killing out." Better still, an explanatory "prologue" can be written to say "we will do no harm with our swords" (3.1.10–17). Prologue can mitigate comparable issues surround-ing the lion, with costume and dialogue serving as the failsafe—let the "lion" wear an ill-fitting mask and speak sweetly, saying, "I would wish you" and "I would request you" (3.1.35). For moonshine, predictive data from an "alma-nac" furnishes the device: "Leave a casement of the great chamber window where we play open " (48–49). Also, let there be a "man in the moon," kitted with a lantern and bush of thorns (3.1.51–52). "Wall" is the other "hard thing," and that is easily fixed by an actor accoutered with "some plaster, or some loam" (3.1.41, 58).

Quince's men are less concerned with holding the mirror up to nature than with preempting operational snags. They like to recycle solutions rather than reinvent the wheel. Their instrumentation relies on objects, materials, elements of nature, and, when nothing else will do, the most complex machine at their disposal—the human body. They would rather avoid illusion and instead ex-hibit "our simple skill"—"wonder on," says the Prologue, "till truth make all things plain" (5.1.110, 127). It does not occur to them that mechanical innova-tion is wondrous, as Guidobaldo insisted: machines can produce miracu-lous spectacles, as when Archimedes single-handedly drew a ship out to sea with a pulley system.[12] In the coming decades, we know, stage engineering would be elevated to a fine art in the court masque; the Vitruvian machin-ery developed by architect Inigo Jones would outdo Ben Jonson's scripts in

---

12. Guidobaldo, *Mechanicorum liber*, 249.

ingenuity.[13] And, as it has been argued, the theater may be thought of as a kind of machine, made of mechanical parts and operated by actor-mechanics, a showcase for technological work.[14] If so, then Shakespeare's artisans show that this machine runs on a particular sort of imagination.

Of course, the players have made a serious miscalculation. While they have exercised their own fantasies, they have overlooked entirely the minds of the spectators, assuming wrongly that the wedding guests have too little imagination to conceive that which is not physically present, and too much imagination not to fear and swoon at what they do see. And yet, however silly the resultant performance might seem, we should note that, from the viewpoint of the thespians, the show goes exactly to plan, with nary a hiccup. As if to show off the fidelity of their execution, the players deictically refer their audience back to the blueprint of their design—the story—throughout the performance. This they present rather than represent ("This man is Pyramus" [5.1.128]) and tell rather than show ("The wall is down" [5.1.338]), intent on pointing out the contractual promises they have kept. "It will fall pat as I told you," says Bottom brightly (5.1.184–85). Predictably, the aristocrats titter. But the reception is not totally hostile. Theseus in particular attempts to decipher the weird representational logic of the production: he notices the mispunctuated prologue, shushes the other guests, and gamely offers that the "man in the moon" ought strictly to be standing inside his lantern (5.1.118, 167, 237–9).

He and Hippolyta pick up their earlier conversation about imagination:

> Theseus. The best in this kind are but shadows, and the worst are no
>   worse if imagination amend them.
> Hippolyta. It must be your imagination, then, and not theirs.
> Theseus. If we imagine no worse of them than they of themselves, they
>   may pass for excellent men. (5.1.208–12)

Perhaps struck by the actors' self-assurance, Theseus remarks on their power of "imagination." Hippolyta reminds him of his duty, as a nobleman and a spectator, to imaginatively supply what is lacking—a familiar sentiment echoed

---

13. Jonson complained that, thanks to Jones's contraptions, "the soul of masque" had become "painting and carpentry." See "An Expostulation with Inigo Jones," in *The Cambridge Edition of the Works of Ben Jonson*, eds. David Bevington, Martin Butler, and Ian Donaldson, vol. 6 (Cambridge: Cambridge University Press, 2012), 50.

14. See Henry S. Turner, "Life Science: Rude Mechanicals, Human Mortals, Posthuman Shakespeare," *South Central Review* 26 (2009): 197–217; and Howard Marchitello, *The Machine in the Text: Science and Literature in the Age of Shakespeare and Galileo* (Oxford: Oxford University Press, 2011).

by many Shakespearean choruses and epilogues, as when Puck begs forgiveness "if we shadows have offended" (5.1.409). Theseus's reply to this sounds like another dig at the delusions of bad performers. It is worded oddly, though, in a way that evokes empathy, not antipathy or even sympathy. In thinking about what these actors "imagine" "of themselves," Theseus seems almost to wish to climb into their minds, to see the theatrical machine of this production from their perspective. "Shadows" is an important choice, since it joins actor, fairy (Oberon is "king of shadows" [3.2.347]), and phantasm, thus coalescing the theatrical, the magical, and the cognitive. In these shadowy lines, Theseus maybe realizes that these not-excellent players are excellent in imagination nonetheless. This might require him to revise his earlier idea of imagination, for it would seem that there is, in addition to the lunatic, lover, and poet, a fourth breed of congenital imaginer—the stage actor.

While it is the earlier speech that is more frequently cited, Shakespeare himself does not privilege the Duke's conception of imagination over the clowns'; if the proportions of the final scene are anything to go by, it is the reverse. The speech is nicely complemented by the performance, at any rate: one is a philosophical disquisition encompassing many elements of imaginative theory; the other is a mechanical demonstration of imagination put into theatrical practice. One spotlights the figure of the inspired poet; the other picks up where the poet—or playwright—leaves off. The design and execution of "Pyramus and Thisbe" takes us through the different imaginative phases of a play's creation that occur after its composition, and it shows us as well the cognitive work that happens on and around the performance stage. Overall, this final act of A Midsummer Night's Dream shows allegorically how theater enforces an expansive and flexible understanding of imagination. Being a man of the theater, it stands to reason that Shakespeare was interested in the practical mechanics of imagination as much as its spontaneous bardic eruptions. He had to be invested in understanding how imagination can work as a workaday instrument of empathetic collective mental transfiguration.

In the later seventeenth and eighteenth centuries, though, this instrumental view would be doubly eclipsed, by a reformulation of Shakespeare's artistic talent and also by new traditions of imaginative theory. Clearly, the Shakespearean fancy had brought to light imagination's importance in aesthetics; however, the literary critics and philosophers of the subsequent era would struggle to reconcile the more mechanical aspects of cognition with the rarefied realms of creativity. During the Enlightenment, A Midsummer Night's Dream was celebrated for its magical, not its mechanical, properties. The play was an exemplification of what in 1712 Joseph Addison, quoting John Dryden, described as Shakespeare's *"Fairy way of Writing"*—a "particular Cast of Fancy, and an Imagination naturally fruitful and superstitious," which "loses sight of Nature" and creates its own "Pattern," producing "Ghosts, Fairies, Witches, and the like Imaginary persons."[15] Whereas Elizabethan poets had once had to defend the virtues of invention, originality and creativity were now prized above mimesis. As surveyed by Judith M. Kennedy and Richard F. Kennedy, the critical tradition of this period found the comedy indisputable proof of the "sublime and amazing imagination, peculiar to Shakespeare, which soars above the bounds of Nature."[16] Stage adaptations boldly remediated the play's imaginative potential with the aid of music, dance, pageantry, and acrobatics.[17] "Pyramus and Thisbe," notably, was excised from the main text and performed separately.[18]

In Addison's wake, imagination expanded into the foundational basis of aesthetic, moral philosophical, and religious experience. To this end, Theseus's speech was the key Shakespearean locus: in Mark Akenside's philosophical poem *The Pleasures of Imagination*, for example, "the child of fancy" is said to be in a "loveliest frenzy caught";

from earth to heav'n he rolls his daring eye,
From heav'n to earth.[19]

Around this time, imagination morphs from a theme in Shakespeare's corpus into an element of his personal sensibility. The trend was begun with John Milton, perhaps, who, reprising the epithet from *Love's Labor's Lost*, christened Shakespeare "fancies childe." Thereafter, Margaret Cavendish notes that *"Shakespear* had a Clear Judgment, a Quick Wit, a Spreading Fancy," and Nahum Tate admires *"Shakespear's* Creating Fancy."[20] Similarly, Addison's ra-

15. Joseph Addison, "The Pleasures of Imagination," in Richard Steele and Joseph Addison, *Selections from The Tatler and The Spectator*, ed. Angus Ross (London: Penguin, 1988), 395.

16. William Warburton, *The Works of Shakespeare* (1747), quoted in *A Midsummer Night's Dream*, ed. Judith M. Kennedy and Richard F. Kennedy, Shakespeare: The Critical Tradition (London: Athlone, 1999), 59. In the same volume, see as well Charles Dibdin, *Complete History of the English Stage* (1800), which dubs the play an exploration of "the regions of fancy" (80), and Nathan Drake, *Shakespeare and his Times* (1817), which finds in it the "fervid and creative power" of Shakespeare's imagination (84).

17. See Jenny Davidson, "Shakespeare Adaptation," in *Shakespeare in the Eighteenth Century*, ed. Fiona Ritchie and Peter Sabor (Cambridge: Cambridge University Press, 2012), 193–96.

18. Louise Geddes explores the afterlife of "Pyramus and Thisbe" in performance in her *Appropriating Shakespeare: A Cultural History of "Pyramus and Thisbe"* (Madison, NJ: Fairleigh Dickinson University Press, 2017).

19. Mark Akenside, *The Pleasures of Imagination*, in *The Poetical Works of Mark Akenside*, ed. Robin Dix (Madison, NJ: Fairleigh Dickinson University Press, 1996), 146.

20. John Milton, "L'Allegro," in *Complete Shorter Poems*, ed. Stella P. Revard (Malden, MA: Wiley-Blackwell, 2009), 52. Margaret Cavendish, *CCXI sociable letters* (London, 1664), 246; Nahum Tate, *The history of King Lear acted at the Duke's theatre* (London, 1681), A2v.

tionale for esteeming Shakespeare above all other English poets is that "noble Extravagance of Fancy, which he had in so great Perfection." Mid-century comes Alexander Gerard's *Essay on Genius*, which ties imagination and also Shakespeare to the concept of genius, which was now something that one was rather than possessed.[21]

Alongside this recharacterization of Shakespeare, imagination was renovated also. By the eighteenth century, the Aristotelian worldview had faded completely, clearing the path for the Newtonian paradigm. Laws of physics provided the primary framework for understanding a universe founded on unvarying mechanical laws. Science, now institutionalized, was focused on experimentation and empirical observation, particularly in those areas that were responsive to these methods of inquiry: mechanics, optics, electricity, magnetism. The theories of cognition that flourished were naturally of a piece with these empirical preoccupations, with little use for *phantasia* as it had been previously conceived, René Descartes now postulated that the faculty had "no part in the abstract certainty of self-reflexive thought" and that the mostly mechanistic workings of the body could be understood without recourse to the soul.[22] The terminological distinctions of faculty psychology, already fluid during the Renaissance, crumbled entirely. Shakespeare, we have seen, explored the discursive landscape of imagination using such nonce words as *phantasime* and *phantasma*. As the seventeenth century progressed, though, the terms *imagination* and *fancy* were all but emptied of significance. Thomas Hobbes calls imagination "decaying sense," suggesting that the difference between sensation, perception, and even memory is only one of intensity, not of kind.[23] John Locke and David Hume similarly elide the distinctions between impressions, images, and ideas, positing instead that the central principle of cognition is that of association among ideas.[24]

The fading of Renaissance psychology probably enabled imagination to come into its own as a natural rather than divine power, to shed its many erstwhile negative associations. Yet eighteenth-century imaginative discourse remains relatively scattered and unsystematic: there did not yet exist an intellectual tradition as robust as the cognitive theoretical system of the scholastics.

21. Addison, "Pleasures," 397; Jonathan Bate, *The Genius of Shakespeare* (New York: Oxford University Press, 1998), 168.

22. Alexander M. Schlutz, *Mind's World: Imagination and Subjectivity from Descartes to Romanticism* (Seattle: University of Washington Press, 2009), 4.

23. Thomas Hobbes, *Leviathan* (London: Penguin, 1968), 88–89.

24. Eva T. H. Brann, *The World of the Imagination: Sum and Substance* (Savage, MD: Rowman and Littlefield, 1991), 80, 83. See as well Richard Olson, "The Human Sciences," 458, and Paul Wood, "Science, Philosophy, and the Mind," 806, both in *Cambridge History of Science*, vol. 4, *Eighteenth-Century Science*, ed. Roy Porter (Cambridge: Cambridge University Press, 2008).

That foundation, James Engell suggests, was laid only when the mechanistic philosophies of the seventeenth century were superseded by ways of thinking that could bestride the empirical and the transcendental.[25] In an influential intervention, Immanuel Kant made imagination a faculty of crucial importance, "the submerged foundation of experience," the place where the a priori synthesis of intuitions and conceptions takes place.[26] As Alexander M. Schlutz points out, the transcendental Kantian imagination is not wholly dissimilar from the Aristotelian *phantasia* in that it occupies a contradictory position and performs a "mediatory function": it is both essential and inferior to reason, "essential to the philosophical method so influential in the development of modern science," while also excluded from it.[27]

When imagination eventually came to be theorized by the British Romantics, the more mechanical elements of the faculty, those very elements that Shakespeare inventively foregrounds with "Pyramus and Thisbe" in *A Midsummer Night's Dream*, were tacitly devalued. William Wordsworth uses Theseus's speech to argue that imagination can never be "merely a faithful copy" of absent objects; rather, it denotes the "processes of creation or of composition." Wordsworth calls the more concrete aspect of image making "Fancy," "given to quicken and to beguile the temporal part of our Nature." By contrast, "Imagination" incites the "eternal," is nothing short of "Reason in her most exalted mood."[28] More famously, Samuel Taylor Coleridge constructs a three-part definition that separates imagination's cognitive, creative, and associative elements. The "prime Agent of all human Perception" is the primary imagination; the poetic power is the "secondary"; and the mechanical amalgamation of memories is fancy.[29] A sensitive reader of Shakespeare, Coleridge esteems him in the first place as a poet, not a dramatist, praising "the perfect sweetness of the versification," the "choice of subjects very remote from the private interests and circumstances of the writer," and those images not "faithfully copied from nature" but rather "modified by a predominant passion, or by associated thoughts or images awakened by that passion."[30] Implicitly di-

---

25. James Engell, *The Creative Imagination: Enlightenment to Romanticism* (Cambridge, MA: Harvard University Press, 1981), 7.

26. Brann, *World of the Imagination*, 89.

27. Schlutz, *Mind's World*, 4, 85.

28. William Wordsworth, preface to *Poems*, 631, 636, and *The Prelude*, 13.170, in *The Major Works*, ed. Stephen Gill (Oxford: Oxford University Press, 2000). See also Joseph M. Ortiz, *Shakespeare and the Culture of Romanticism* (Farnham, England: Ashgate, 2013), 90.

29. Samuel Taylor Coleridge, *Biographia Literaria*, vol. 1, in *The Collected Works of Samuel Taylor Coleridge*, ed. James Engell and Walter Jackson Bate (London: Routledge and Kegan Paul, 1983), 7:304–5.

30. Samuel Taylor Coleridge, *Coleridge on Shakespeare*, ed. Terence Hawkes (Harmondsworth, England: Penguin, 1969), 68–71. See as well Charles Mahoney, "Coleridge and Shakespeare," in *The*

minished in this characterization are the physicality and technicity of the stage; the collaborative nature of theater; the imaginative labor performed by audiences and actors in addition to the author.

After Coleridge, the still-fulsome critical praise of *A Midsummer Night's Dream* is edged with dissatisfaction about the theater. "This play is not for the stage," writes one critic; echoing Hippolyta, he adds, "Our imagination must amend what is wanting." Another critic hears in the moment when Theseus likens actors to "shadows" Shakespeare griping about "the common calamity of dramatists"—namely, their enforced reliance on the "mind of the spectator."[31] Just as Theseus's speech on the lunatic, lover, and poet had earlier been an authoritative Shakespearean statement about creativity, Theseus's remark about shadowy actors becomes a weary Shakespearean confession about the limitations of his art.[32] William Hazlitt is especially emphatic in his observation that whenever *A Midsummer Night's Dream* is acted on stage, "all that is finest in the play," its "spirit" and "genius," flees. He extrapolates this into a generalized statement about the incompatibility of imagination and theater: "Fancy cannot be embodied any more than a simile can be painted; and it is as idle to attempt it as to personate *Wall* or *Moonshine*."[33] According to Hazlitt, the failure of "Pyramus and Thisbe" is not so much anomalous as typical, illustrative of the inadequacies of theater.

All this considered, it is surely no exaggeration to say that the critical reception of Shakespeare has in itself played a role in the history of imagination. Clearly, *A Midsummer Night's Dream* spurred writers and thinkers to salvage imagination's aesthetic and intellectual fate. But to do this meant in practice to wrest Shakespeare from the theater; this may in turn have obscured our view of the epistemological problems that originally interested him. What also seems clear is that Shakespeare's moment in history was singular. It was a moment when premodern psychology was stalled and imagination was gaining prominence in the context of poetics; it was a period of paradigmatic change, disciplinary reorganization, and general intellectual and epistemic expansion; finally, it was the early days of the institutionalized theater, the public and private playhouses. It was still possible to believe that phantasms are forged in cavities of the brain, possible to ponder fancy's potential role in scientific dis-

Oxford Handbook of Samuel Taylor Coleridge, ed. Frederick Burwick (Oxford: Oxford University Press, 2009), 504, 513.

31. Charles Knight, *The Pictorial Edition of Shakespeare's Works* (1839), quoted in Kennedy and Kennedy, *Midsummer*, 127–28; William Maginn, from "Shakespeare Papers.—No. IV. Midsummer Night's Dream. Bottom, the Weaver" (1837), quoted in Kennedy and Kennedy, *Midsummer*, 113.

32. David Fairer, "Shakespeare in Poetry," in Ritchie and Sabor, *Shakespeare*, 103.

33. William Hazlitt, *Characters of Shakespeare's Plays* (London: Oxford University Press, 1955), 103.

covery, possible to think of *scientia* and *ars* completely interchangeably. It would not be so for long. The pre-Cartesian Renaissance is much too often glossed over in the history of imagination, even though the period produced a literary practitioner who greatly contributed to imagination's enduring importance. We need a new idea of genius, a new way of describing what Shakespeare achieved, which is as much concerned with transdisciplinary curiosity and problem solving as with inspiration and philosophy. What the Coleridgean imagination would unify into an "esemplastic" whole, the Shakespearean one leaves multifarious, a discord that must be shaped into concord with all the agential and intellective resources that human beings can muster.

This task falls to Bottom when he wakens from his enchanted slumber. In a funny yet also moving monologue, he tries to describe what he has seen: "I have had a most rare vision. I have had a dream past the wit of man to say what dream it was. Man is but an ass if he go about to expound this dream. . . . The eye of man hath not heard, the ear of man hath not seen, man's hand is not able to taste, his tongue to conceive, nor his heart to report what my dream was" (4.1.202–10). As is usually noted, these lines are a garbled quotation of Paul. They seem to present an imaginative epiphany, an oneiric transport handed down from above. Yet the speech is undercut by heavy dramatic irony, by our knowing very well who has been the real "ass." If there is sublimity here, it draws less from the content of the vision, perhaps, than from the halting suppositions of the seer. This is a moment that repeats throughout Shakespeare, where we find many kinds of men and women—courtiers and explorers, young lovers and aging kings—interrogating not only their perceptions but the very idea of perception. Evidently, Shakespeare understood that this witnessing of mentality, wherein we watch other people reflect on the mind's machinery, is inexplicably rapturous in its way, frequently moving. The reflection does not have to be fruitful or even very coherent to produce the effect, as Bottom's comic synesthesia demonstrates. It does depend on the fusion between his inconclusive cognitive parsing of the dream and our parsing of him—the special fusion that happens only and always at the theater.

The notion of numinous inspiration is also being mocked here, for even though Bottom can neither "conceive" nor "report" his dream, he is blithely confident of its poetic value: "I will get Peter Quince to write a ballad of this dream. It shall be called 'Bottom's Dream,' because it hath no bottom; and I will sing it in the latter end of a play, before the Duke" (4.1.210–13). How is he going to render into poetry what he cannot even put into words? He may do it, perhaps, in a "fine frenzy," like Theseus's poet. Moments later, however, the would-be balladeer transforms into an actor in search of his company. If we look at the entire speech, in fact, we see that Bottom's epiphanic vision is bracketed at both ends by the lure of the stage—which functions here almost as a grounding metonym for the earthly, human world. Opening his eyes, Bottom's first words are not those of a prophet; he sounds more like a restive actor backstage: "When my cue comes, call me, and I will answer" (4.1.198–99). His initial thought is not that "man is but an ass"; it is to rejoin his fellow men, whose names and vocations he utters reflexively: "Peter Quince? Flute the bellows-mender? Snout the tinker? Starveling?" (4.1.199–201). And, as the fairy revelation recedes and his mind veers back toward the theatrical task at hand, he revises his initial poetic impulse. "Bottom's Dream," Bottom thinks, might be best of all served if it were sung before a live audience—maybe "in the latter end of a play" (4.1.212). For now it will have to wait, at any rate, because another play is about to begin.

BIBLIOGRAPHY

Primary Works

Agrippa, Henry Cornelius. *Of the Vanitie and Vncertaintie of Artes and Sciences.* Edited by Catherine M. Dunn. Northridge: California State University, 1974.

Akenside, Mark. *The Poetical Works of Mark Akenside,* edited by Robin Dix. Madison, NJ: Fairleigh Dickinson University Press, 1996.

Aquinas, Thomas. *On Human Nature.* Edited by Thomas S. Hibbs. Indianapolis: Hackett, 1999.

———. *Summa Theologiæ.* Vol. 11. Translated by Timothy Suttor. Cambridge: Cambridge University Press, 2006.

Aristotle. *Metaphysics, Volume I: Books 1–9.* Translated by Hugh Tredennick. Loeb Classical Library 271. Cambridge, MA: Harvard University Press, 1933.

———. *On the Soul. Parva Naturalia. On Breath.* Translated by W. S. Hett. Loeb Classical Library 288. Cambridge, MA: Harvard University Press, 1957.

———. *Parts of Animals. Movement of Animals. Progression of Animals.* Translated by A. L. Peck, E. S. Forster. Loeb Classical Library 323. Cambridge, MA: Harvard University Press, 1937.

———. *Problems, Volume II: Books 20–38. Rhetoric to Alexander.* Translated and edited by Robert Mayhew and David C. Mirhady. Loeb Classical Library 317. Cambridge, MA: Harvard University Press, 2011.

Armenini, Giovanni Battista. *On the True Precepts of the Art of Painting.* Edited by Edward J. Olszewski. New York: Burt Franklin, 1977.

Augustine. *Confessions, Volume I: Books 1–8.* Translated by Carolyn J.-B. Hammond. Loeb Classical Library 26. Cambridge, MA: Harvard University Press, 2014.

———. *On the Trinity.* Edited by Gareth B. Matthews. Translated by Stephen McKenna. Cambridge: Cambridge University Press, 2002.

Bacon, Francis. *The Works of Francis Bacon.* Edited by James Spedding, Robert Leslie Ellis, and Douglas Denon Heath. 14 vols. London: Longman, 1857–1874.

Bailey, Walter. *Two treatises concerning the preseruation of eie-sight.* Oxford, 1616.

Barnes, Barnabe. *Parthenophil and Parthenophe.* London, 1593.

Barrough, Philip. *The methode of phisicke.* London, 1583.

Batman, Stephen. *Batman vppon Bartholome his booke De proprietatibus rerum.* London, 1582.

Beard, Thomas. *A retractiue from the Romish religion.* London, 1616.

Bevington, David, Lars Engle, Katharine Eisaman Maus, and Eric Rasmussen, eds. *English Renaissance Drama.* New York: W. W. Norton, 2008.

The Bible: Authorized King James Version with Apocrypha. Edited by Robert Carroll and
     Stephen Prickett. Oxford: Oxford University Press, 1997.
Boaistuau, Pierre. *Certaine secrete wonders of nature.* Translated by Edward Fenton.
     London, 1569.
Boorde, Andrew. *The Breuiary of helthe.* London, 1547.
Breton, Nicholas. *Brittons bowre of delights.* London, 1591.
———. *A floorish vpon fancie.* London, 1577.
Bright, Timothy. *A treatise of melancholie.* London, 1586.
Bull, Henry. *Christian praiers and holie meditations.* London, 1578.
Burton, Robert. *The Anatomy of Melancholy.* Edited by Thomas C. Faulkner,
     Nicolas K. Kiessling, and Rhonda L. Blair. 6 vols. Oxford: Clarendon, 1989.
Calvin, John. *The institution of Christian religion.* Translated by Thomas Norton.
     London, 1561.
———. *Sermons of Master Iohn Calvin, vpon the booke of Iob.* Translated by Arthur
     Golding. London, 1574.
Cardano, Girolamo. *The "De Subtilitate" of Girolamo Cardano.* Edited by John M.
     Forrester. 2 vols. Tempe: Arizona Center for Medieval and Renaissance
     Studies, 2013.
Cartwright, Thomas. *An hospitall for the diseased.* London, 1580.
Cavendish, Margaret. *CCXI sociable letters.* London, 1664.
———. *Observations upon experimental philosophy.* London, 1666.
———. *Poems, and fancies.* London, 1653.
Charron, Pierre. *Of wisdome.* Translated by Samson Lennard. London, 1608.
Coenen, Adriaen. *The Whale Book: Whales and Other Marine Animals as Described by
     Adriaen Coenen in 1585.* Edited by Florike Egmond and Peter Mason. London:
     Reaktion, 2003.
Coleridge, Samuel Taylor. *Coleridge on Shakespeare.* Edited by Terence Hawkes.
     Harmondsworth, England: Penguin, 1969.
———. *The Collected Works of Samuel Taylor Coleridge.* Edited by James Engell and
     Walter Jackson Bate. Vol. 7. London: Routledge and Kegan Paul, 1983.
Constable, Henry. *Diana.* London, 1592.
Cooper, Thomas. *Thesaurus linguae Romanae & Britannicae.* London, 1578.
Cornwallis, William. *Essayes.* London, 1600.
Cranmer, Thomas. *An aunswere by the Reuerend Father in God Thomas Archbyshop of
     Canterbury.* London, 1580.
———. *A defence of the true and catholike doctrine of the sacrament of the body and bloud of
     our sauiour Christ.* London, 1550.
Crooke, Helkiah. *Mikrokosmographia.* London, 1615.
Daneau, Lambert. *A dialogue of witches.* London, 1575.
Davies, John. *The Gulling Sonnets.* New York: Columbia University Press, 1941.
———. *Nosce teipsum.* London, 1599.
Davies, John, of Hereford. *Microcosmos.* Oxford, 1603.
———. *Mirum in modum.* London, 1602.
———. *The Muses sacrifice.* London, 1612.
Deacon, John. *Dialogicall discourses of spirits and divels.* London, 1601.
Dekker, Thomas. *The pleasant comedie of old Fortunatus.* London, 1600.

Descartes, René. *Discourse of Method, Optics, Geometry, and Meteorology.* Translated by Paul J. Olscamp. Indianapolis: Hackett, 2001.

Digby, Kenelm. *Two treatises in the one of which the nature of bodies, in the other, the nature of mans soule is looked into.* Paris, 1644.

Drake, Stillman, and I. E. Drabkin, trans. *Mechanics in Sixteenth-Century Italy.* Madison: University of Wisconsin Press, 1969.

Drayton, Michael. *Englands heroicall epistles.* London, 1599.

Du Bartas, Guillaume de Salluste. *Du Bartas his deuine weekes and workes translated.* Translated by Josuah Sylvester. London, 1611.

Du Laurens, André. *A discourse of the preseruation of the sight.* Translated by Richard Surphlet. London, 1599.

Edwards, Richard. *The paradise of daintie deuises.* London, 1585.

Elyot, Thomas. *The castel of helth.* London, 1539.

——. *The dictionary of syr Thomas Eliot knyght.* London, 1538.

Epicurus. *The Epicurus Reader: Selected Writings and Testimonia.* Translated and edited by Brad Inwood and L. P. Gerson. Indianapolis: Hackett, 1994.

Erasmus, Desiderius. *The first tome or volume of the Paraphrase of Erasmus vpon the newe testamente.* Translated by Nicholas Udall. London, 1548.

Estienne, Charles. *The defence of contraries.* Translated by Anthony Munday. London, 1593.

Euclid. *The elements of geometrie.* Translated by Henry Billingsley. London, 1570.

——. "The Optics of Euclid." Translated by Harry Edwin Burton. *Journal of the Optical Society of America* 35, no. 5 (1945): 357–72.

Fernel, Jean. *Jean Fernel's "On the Hidden Causes of Things": Forms, Souls, and Occult Diseases in Renaissance Medicine.* Edited by John M. Forrester. Leiden, Netherlands: Brill, 2005.

Ficino, Marsilio. *Platonic Theology.* Vol. 4, translated by Michael J. B. Allen, edited by James Hankins. Cambridge, MA: Harvard University Press, 2004.

——. *Three Books on Life.* Translated and edited by Carol V. Kaske and John R. Clark. Binghamton, NY: Medieval & Renaissance Texts & Studies in conjunction with the Renaissance Society of America, 1989.

Fracastoro, Girolamo. *Hieronymi Fracastorii De contagione et contagiosis morbis et eorum curatione.* Translated by Wilmer Cave Wright. New York: G. P. Putnam's Sons, 1930.

Fraunce, Abraham. *The lawiers logike.* London, 1588.

Galen. *Galen: Selected Works.* Translated by P. N. Singer. Oxford: Oxford University Press, 1997.

——. *On the Affected Parts.* Translated and edited by Rudolph E. Siegel. Basel, Switzerland: Karger, 1976.

Gassendi, Pierre. *The Selected Works of Pierre Gassendi.* Translated and edited by Craig B. Bush. New York: Johnson Reprint, 1972.

The Geneva Bible: A Facsimile of the 1599 Edition. Ozark, MO: L. L. Brown, 1990.

Gifford, George. *A dialogue concerning witches and witchcraftes.* London, 1593.

Gilbert, Allan H., ed. *Literary Criticism: Plato to Dryden.* New York: American Book Company, 1940.

Greville, Fulke. *Certaine learned and elegant workes.* London, 1633.

Guillemeau, Jacques. *A treatise of one hundred and thirteene diseases of the eyes.* Translated by Richard Banister. London, 1622.

Hakewill, George. *The vanitie of the eye.* Oxford, 1608.

Hazlitt, William. *Characters of Shakespeare's Plays.* London: Oxford University Press, 1955.

Hill, Thomas. *The most pleasaunte Arte of the interpretacion of Dreames.* London, 1576.

Hobbes, Thomas. *Leviathan.* London: Penguin, 1968.

Holland, Henry. *A treatise against witchcraft.* Cambridge, England, 1590.

Huarte, Juan. *The examination of mens wits.* Translated by Richard Carew. London, 1594.

Hugh of Saint Victor. *The Didascalicon of Hugh of Saint Victor.* Translated by Jerome Taylor. New York: Columbia University Press, 1991.

Jewel, John. *A replie vnto M. Hardinges answeare.* London, 1565.

Jones, John. *The arte and science of preseruing bodie and soule.* London, 1579.

Jonson, Ben. *The Cambridge Edition of the Works of Ben Jonson.* Edited by David Bevington, Martin Butler, and Ian Donaldson. 7 vols. Cambridge: Cambridge University Press, 2012.

Jorden, Edward. *A briefe discourse of a disease called the suffocation of the mother.* London, 1603.

Junius, Franciscus. *The painting of the ancients.* London, 1638.

Kepler, Johannes. *Gesammelte Werke.* Edited by Walther von Dyck and Max Caspar. 19 vols. Munich: C. H. Beck, 1938.

——. *Optics.* Translated by William H. Donahue. Santa Fe: Green Lion, 2000.

Langton, Christopher. *An introduction into phisycke.* London, 1545.

La Primaudaye, Pierre de. *The French academie.* Translated by Thomas Bowes. London, 1586.

——. *The second part of the French academie.* Translated by Thomas Bowes. London, 1594.

Lavater, Ludwig. *Of ghostes and spirites walking by nyght.* Translated by Robert Harrison. London, 1572.

Le Loyer, Pierre. *A treatise of specters or straunge sights.* Translated by Zachary Jones. London, 1605.

Lemnius, Levinus. *The touchstone of complexions.* Translated by Thomas Newton. London, 1576.

Leo, John. *A geographical historie of Africa.* Translated by John Pory. London, 1600.

Leonardo da Vinci. *Leonardo on Painting.* Translated by Martin Kemp and Margaret Walker, edited by Martin Kemp. New Haven, CT: Yale University Press, 1989.

——. *Leonardo's Notebooks: Writing and Art of the Great Master.* Edited by H. Anna Suh. New York: Black Dog and Leventhal, 2013.

Léry, Jean de. *History of a Voyage to the Land of Brazil.* Translated by Janet Whatley. Berkeley: University of California Press, 1990.

Lomazzo, Giovan Paolo. *Idea of the Temple of Painting.* Translated and edited by Jean Julia Chai. University Park: Pennsylvania State University Press, 2013.

Longinus. *On the Sublime.* Translated by W. Hamilton Fyfe. Revised by Donald Russell. In Aristotle, Longinus, Demetrius, *Poetics. Longinus: On the Sublime. Demetrius: On Style.* Translated by Stephen Halliwell, W. Hamilton Fyfe,

Doreen C. Innes, W. Rhys Roberts. Revised by Donald A. Russell. Loeb
    Classical Library 199. Cambridge, MA: Harvard University Press, 1995.
Lucretius. *On the Nature of Things*. Translated by W. H. D. Rouse. Revised by
    Martin F. Smith. Loeb Classical Library 181. Cambridge, MA: Harvard Univer-
    sity Press, 2002.
Mackay, Christopher S., trans. and ed. *The Hammer of Witches*. Cambridge: Cam-
    bridge University Press, 2009.
Mandeville, John. *The Travels of Sir John Mandeville*. Edited by E. C. Coleman. Stroud,
    England: Nonsuch, 2006.
Marnix van St. Aldegonde, Philips van. *The bee hiue of the Romishe Church*. London,
    1579.
Milton, John. *Complete Shorter Poems*. Edited by Stella P. Revard. Malden, MA:
    Wiley-Blackwell, 2009.
Montaigne, Michel de. *The essayes*. Translated by John Florio. London, 1603.
Moore, Philip. *The hope of health*. London, 1565.
Mornay, Philippe de, seigneur du Plessis-Marly. *A woorke concerning the trewnesse of
    the Christian religion*. Translated by Philip Sidney and Arthur Golding. London,
    1587.
Munday, Anthony. *A banquet of daintie conceits*. London, 1588.
Nash, Thomas. *The terrors of the night*. London, 1594.
Ovid. *The .xv. bookes of P. Ouidius Naso, entytuled Metamorphosis*. Translated by Arthur
    Golding. London, 1567.
Paré, Ambroise. *The workes of that famous chirurgion Ambrose Parey*. Translated by
    Thomas Johnson. London, 1634.
Perkins, William. *The first part of The cases of conscience*. London, 1604.
——. *A reformed Catholike*. Cambridge, England, 1598.
——. *The whole treatise of the cases of conscience*. London, 1606.
Petrarch. *Petrarch's Lyric Poems*. Translated and edited by Robert Durling. Cam-
    bridge, MA: Harvard University Press, 1976.
Philostratus. *Apollonius of Tyana, Volume I: Life of Apollonius of Tyana, Books 1–4*.
    Translated and edited by Christopher P. Jones. Loeb Classical Library 16.
    Cambridge, MA: Harvard University Press, 2005.
Pico della Mirandola, Gianfrancesco. *On the Imagination*. Translated and edited by
    Harry Caplan. New Haven, CT: Yale University Press, 1930.
Plato. *Theaetetus. Sophist*. Translated by Harold North Fowler. Loeb Classical Library
    123. Cambridge, MA: Harvard University Press, 1967.
——. *Timaeus. Critias. Cleitophon. Menexenus. Epistles*. Translated by R. G. Bury. Loeb
    Classical Library 234. Cambridge, MA: Harvard University Press, 1929.
Pliny. *The historie of the world*. Translated by Philemon Holland. London, 1601.
——. *Natural History, Volume II: Books 3–7*. Translated by H. Rackham. Cambridge,
    MA: Harvard University Press, 1942.
Plotinus. *Ennead, Volume I: Porphyry on the Life of Plotinus. Ennead I*. Translated by
    A. H. Armstrong. Loeb Classical Library 440. Cambridge, MA: Harvard
    University Press, 1969.
——. *Ennead, Volume IV*. Translated by A. H. Armstrong. Loeb Classical Library 443.
    Cambridge, MA: Harvard University Press, 1984.

Plutarch. *The philosophie, commonlie called, the morals.* Translated by Philemon Holland. London, 1603.

Porta, Giambattista della. *Natural magick.* London, 1658.

Ptolemy. *Ptolemy's Theory of Visual Perception: An English Translation of the "Optics."* Translated by A. Mark Smith. Independence Square, PA: American Philosophical Society, 1996.

Purchas, Samuel. *Purchas his pilgrimage.* London, 1613.

Puttenham, George. *The Art of English Poesy.* Edited by Frank Whigham and Wayne A. Rebhorn. Ithaca, NY: Cornell University Press, 2007.

Quintilian. *The Orator's Education, Volume III: Books 6–8.* Translated and edited by Donald A. Russell. Loeb Classical Library 126. Cambridge, MA: Harvard University Press, 2001.

Radden, Jennifer, ed. *The Nature of Melancholy: From Aristotle to Kristeva.* Oxford: Oxford University Press, 2000.

Rainolds, William. *A treatise conteyning the true catholike and apostolike faith of the holy sacrifice and sacrament ordeyned by Christ at his last Supper.* Antwerp, 1593.

Raleigh, Walter. *The history of the world.* London, 1614.

Sander, Nicholas. *The supper of our Lord.* Leuven, Belgium, 1565.

Savonarola, Girolamo. *Selected Writings of Girolamo Savonarola: Religion and Politics, 1490–1498.* Translated and edited by Anne Borelli and Maria Pastore Passaro. New Haven, CT: Yale University Press, 2006.

Scot, Reginald. *The discouerie of witchcraft.* London, 1584.

Scribonius, Wilhelm Adolf. *Naturall philosophy.* Translated by Daniel Widdowes. London, 1621.

Sennert, Daniel. *The institutions or fundamentals of the whole art, both of physick and chirurgery.* London, 1656.

——. *Thirteen books of natural philosophy.* London, 1660.

Shakespeare, William. *The Norton Shakespeare.* 3rd ed. Edited by Stephen Greenblatt, Walter Cohen, Suzanne Gossett, Jean E. Howard, Katharine Eisaman Maus, and Gordon McMullan. New York: W. W. Norton, 2016.

Sidney, Philip. *The Major Works.* Edited by Katherine Duncan-Jones. Oxford: Oxford University Press, 2009.

Spenser, Edmund. *Amoretti and Epithalamion.* Edited by Kenneth J. Larsen. Medieval and Renaissance Texts and Studies 146. Tempe: Arizona State University, 1997.

——. *The Faerie Qveene.* Edited by A. C. Hamilton, text edited by Hiroshi Yamashita and Toshiyuki Suzuki. New York: Longman, 2001.

Steele, Richard, and Joseph Addison. *Selections from The Tatler and The Spectator.* Edited by Angus Ross. London: Penguin, 1988.

Tasso, Torquato. *Discourses on the Heroic Poem.* Translated by Mariella Cavalchini and Irene Samuel. Oxford: Clarendon, 1973.

Tate, Nahum. *The history of King Lear acted at the Duke's theatre.* London, 1681.

Tofte, Robert. *Alba.* London, 1598.

——. *Laura.* London, 1597.

Topsell, Edward. *The historie of foure-footed beastes.* London, 1607.

——. *The historie of serpents.* London, 1608.

Tyndale, William. *The whole workes of W. Tyndall.* London, 1573.

Valverde de Amusco, Juan. *Anatomia del corpo humano*. Rome, 1559.

Vasari, Giorgio. *The Lives of Painters, Sculptors and Architects*. Edited by William Gaunt. 4 vols. London: Dent, 1963.

Vermigli, Peter Martyr. *A discourse or traictise of Petur Martyr Vermilla Florentine*. Translated by Nicholas Udall. London, 1550.

Vesalius, Andreas. *De humani corporis fabrica libri septem*. Brussels: Culture and Civilization, 1964.

——. *The Fabric of the Human Body*. Translated and edited by Daniel H. Garrison and Malcolm H. Hast. 2 vols. Basel, Switzerland: Karger, 2014.

——. *The Illustrations from the Works of Andreas Vesalius of Brussels*. Edited by J. B. de C. M. Saunders and Charles D. O'Malley. Cleveland: World, 1950.

Vicary, Thomas. *The Englishemans treasure*. London, 1587.

——. *A profitable treatise of the anatomie of mans body*. London, 1577.

Watson, Thomas. *The tears of fancie*. London, 1593.

Wordsworth, William. *The Major Works*. Edited by Stephen Gill. Oxford: Oxford University Press, 2000.

Wright, Thomas. *The passions of the minde in generall*. London, 1604.

Secondary Works

Acheson, Katherine. "Gesner, Topsell, and the Purposes of Pictures in Early Modern Natural Histories." In *Printed Images in Early Modern Britain: Essays in Interpretation*, edited by Michael Hunter, 127–44. Farnham, England: Ashgate, 2010.

Alpers, Paul. "*King Lear* and the Theory of the 'Sight Pattern.'" In *In Defense of Reading: A Reader's Approach to Literary Criticism*, edited by Reuben A. Brower and Richard Poirier, 133–52. New York: E. P. Button, 1962.

Alpers, Svetlana. *The Art of Describing: Dutch Art in the Seventeenth Century*. London: Murray, 1983.

Anderson, Judith H., and Jennifer C. Vaught, eds. *Shakespeare and Donne: Generic Hybrids and the Cultural Imaginary*. New York: Fordham University Press, 2013.

Archer, John Michael. "*Love's Labour's Lost*." In *A Companion to Shakespeare's Works*, edited by Richard Dutton and Jean E. Howard, 3:320–37. Oxford: Blackwell, 2003.

Ashworth, William B., Jr. "Emblematic Natural History of the Renaissance." In *Cultures of Natural History*, edited by N. Jardine, J. A. Secord, and E. C. Spary, 17–37. Cambridge: Cambridge University Press, 1996.

——. "Natural History and the Emblematic World View." In *Reappraisals of the Scientific Revolution*, edited by David C. Lindberg and Robert S. Westman, 303–32. Cambridge: Cambridge University Press, 1990.

Babb, Lawrence. *The Elizabethan Malady: A Study of Melancholia in English Literature from 1580 to 1642*. East Lansing: Michigan State College Press, 1951.

Bambach, Carmen C., ed. *Leonardo da Vinci: Master Draftsman*. New Haven, CT: Yale University Press, 2003.

Barber, C. L. *Shakespeare's Festive Comedy*. Princeton, NJ: Princeton University Press, 1972.

Bate, Jonathan. *The Genius of Shakespeare*. New York: Oxford University Press, 1998.

Belsey, Catherine. "The Name of the Rose in *Romeo and Juliet.*" *Yearbook of English Studies* 23 (1993): 126–42.

Berger, Harry. *Situated Utterances: Texts, Bodies, and Cultural Representations.* New York: Fordham University Press, 2005.

Bergeron, David. "Sickness in *Romeo and Juliet.*" *College Language Association Journal* 20 (1977): 356–64.

Berthelot, Joseph A. *Michael Drayton.* New York: Twayne, 1967.

Biagioli, Mario. "Etiquette, Interdependence, and Sociability in Seventeenth-Century Science." *Critical Inquiry* 22, no. 2 (1996): 193–238.

Boehrer, Bruce. *Shakespeare among the Animals: Nature and Society in the Drama of Early Modern England.* New York: Palgrave, 2002.

Booth, Stephen, ed. *Shakespeare's Sonnets.* New Haven, CT: Yale University Press, 1977.

Bozio, Andrew. "Embodied Thought and the Perception of Place in *King Lear.*" *SEL Studies in English Literature 1500–1900* 55, no. 2 (2015): 263–284.

Bradley, A. C. *Shakespearean Tragedy.* London: Penguin, 1991.

Brann, Eva T. H. *The World of the Imagination: Sum and Substance.* Savage, MD: Rowman and Littlefield, 1991.

Brann, Noel L. *The Debate over the Origin of Genius during the Italian Renaissance.* Leiden, Netherlands: Brill, 2002.

Brown, Alison. *The Return of Lucretius to Renaissance Florence.* Cambridge, MA: Harvard University Press, 2010.

Brown, John Russell, ed. *Focus on "Macbeth."* London: Routledge, 1982.

Bruster, Douglas. *Quoting Shakespeare: Form and Culture in Early Modern Drama.* Lincoln: University of Nebraska Press, 2000.

Bullough, Geoffrey, ed. *Narrative and Dramatic Sources of Shakespeare.* 8 vols. London: Routledge, 1957–1975.

Bundy, Murray Wright. "The Allegory in *The Tempest.*" *Research Studies* 32 (1964): 189–206.

——. *The Theory of Imagination in Classical and Medieval Thought.* Urbana: University of Illinois Press, 1927.

Burnett, Mark Thornton. *Constructing "Monsters" in Shakespearean Drama and Early Modern Culture.* Basingstoke, England: Palgrave Macmillan, 2002.

Calderwood, James. *If It Were Done: "Macbeth" and Tragic Action.* Amherst: University of Massachusetts Press, 1986.

Cantelupe, Eugene B. "An Iconographical Interpretation of *Venus and Adonis,* Shakespeare's Ovidian Comedy." *Shakespeare Quarterly* 14, no. 2 (1963): 141–51.

Cartelli, Thomas. "Banquo's Ghost: The Shared Vision." *Theatre Journal* 35, no. 3 (1983): 389–405.

Clark, Stuart. *Thinking with Demons: The Idea of Witchcraft in Early Modern Europe.* Oxford: Oxford University Press, 1997.

——. *Vanities of the Eye: Vision in Early Modern European Culture.* Oxford: Oxford University Press, 2007.

Cocking, J. M. *Imagination: A Study in the History of Ideas.* Edited by Penelope Murray. London: Routledge, 1991.

Cohen, Stephen. *Shakespeare and Historical Formalism*. Aldershot, England: Ashgate, 2007.

Colie, Rosalie. *Paradoxia Epidemica: The Renaissance Tradition of Paradox*. Princeton, NJ: Princeton University Press, 1966.

——. *Shakespeare's Living Art*. Princeton, NJ: Princeton University Press, 1974.

Corum, Richard. "'The Catastrophe Is a Nuptial': *Love's Labor's Lost*, Tactics, Everyday Life." In *Renaissance Culture and the Everyday*, edited by Patricia Fumerton and Simon Hunt, 271–98. Philadelphia: University of Pennsylvania Press, 1999.

Crane, Mary Thomas. *Losing Touch with Nature: Literature and the New Science in Sixteenth-Century England*. Baltimore: Johns Hopkins University Press, 2014.

——. *Shakespeare's Brain: Reading with Cognitive Theory*. Princeton, NJ: Princeton University Press, 2001.

Crombie, A. C. *Augustine to Galileo: The History of Science, A.D. 400–1650*. London: Falcon, 1952.

——. "Early Concepts of the Senses and the Mind." *Scientific American* 210 (1964): 24–116.

Cummins, Juliet, and David Burchell, eds. *Science, Literature and Rhetoric in Early Modern England*. Aldershot, England: Ashgate, 2007.

Cunningham, Andrew. *The Anatomical Renaissance: The Resurrection of the Anatomical Projects of the Ancients*. Aldershot, England: Scolar, 1997.

Curran, Kevin. "Feeling Criminal in *Macbeth*." *Criticism* 54, no. 3 (2012): 391–401.

Darrigol, Olivier. *A History of Optics: From Greek Antiquity to the Nineteenth Century*. Oxford: Oxford University Press, 2012.

Daston, Lorraine. "Baconian Facts, Academic Civility, and the Prehistory of Objectivity." *Annals of Scholarship* 8 (1991): 337–63.

——. "Marvelous Facts and Miraculous Evidence in Early Modern Europe." In *Wonders, Marvels, and Monsters in Early Modern Culture*, edited by Peter G. Platt, 76–104. Newark: University of Delaware Press, 1999.

Daston, Lorraine, and Katharine Park. *Wonders and the Order of Nature, 1150–1750*. Cambridge, MA: MIT Press, 2001.

Dear, Peter. *Discipline and Experience: The Mathematical Way in the Scientific Revolution*. Chicago: University of Chicago Press, 1995.

——. *Revolutionizing the Sciences: European Knowledge and Its Ambitions, 1500–1700*. Princeton, NJ: Princeton University Press, 2009.

——, ed. *The Scientific Enterprise in Early Modern Europe: Readings from "Isis."* Chicago: University of Chicago Press, 1997.

Diehl, Houston. "Horrid Image, Sorry Sight, Fatal Vision: The Visual Rhetoric of *Macbeth*." *Shakespeare Studies* 16 (1983): 191–203.

Dubrow, Heather. *A Happier Eden: The Politics of Marriage in the Stuart Epithalamion*. Ithaca, NY: Cornell University Press, 1990.

Duncan-Jones, Katherine, ed. *Shakespeare's Sonnets*. London: Thomas Nelson, 1997.

Eamon, William. "Court, Academy, and Printing House: Patronage and Scientific Careers in Late-Renaissance Italy." In *Patronage and Institutions: Science, Technology, and Medicine at the European Court, 1500–1750*, edited by Bruce T. Moran, 25–50. Rochester, NY: Boydell, 1991.

——. *Science and the Secrets of Nature: Books of Secrets in Medieval and Early Modern Culture*. Princeton, NJ: Princeton University Press, 1994.

Edgerton, Samuel Y., Jr. *The Renaissance Rediscovery of Linear Perspective*. New York: Harper and Row, 1976.

Elliott, J. H. *The Old World and the New, 1492–1650*. Cambridge: Cambridge University Press, 1970.

Emerton, Norma E. *The Scientific Reinterpretation of Form*. Ithaca, NY: Cornell University Press, 1984.

Engell, James. *The Creative Imagination: Enlightenment to Romanticism*. Cambridge, MA: Harvard University Press, 1981.

Fineman, Joel. *Shakespeare's Perjured Eye: The Invention of Poetic Subjectivity in the Sonnets*. Berkeley: University of California Press, 1986.

Floyd-Wilson, Mary. *Occult Knowledge, Science, and Gender on the Shakespearean Stage*. Cambridge: Cambridge University Press, 2013.

Floyd-Wilson, Mary, and Garrett A. Sullivan Jr., eds. *Environment and Embodiment in Early Modern England*. Basingstoke, England: Palgrave Macmillan, 2007.

Forest, Louise C. Turner. "A Caveat for Critics against Invoking Elizabethan Psychology." *PMLA* 61, no. 3 (1946): 651–72.

Foucault, Michel. *The Archaeology of Knowledge*. Translated by A. M. Sheridan. London: Routledge, 2002.

——. *The Order of Things: An Archaeology of the Human Sciences*. New York: Pantheon, 1970.

Frye, Northrop. *Northrop Frye's Writings on Shakespeare and the Renaissance*. Edited by Troni Y. Grande and Garry Sherbert. Toronto: University of Toronto Press, 2010.

Gallagher, Lowell, and Shankar Raman, eds. *Knowing Shakespeare: Senses, Embodiment and Cognition*. Basingstoke, England: Palgrave Macmillan, 2010.

Garber, Marjorie. *Dream in Shakespeare: From Metaphor to Metamorphosis*. New Haven, CT: Yale University Press, 1974.

——. *Shakespeare After All*. New York: Pantheon, 2004.

Garnier-Giamarchi, Marie. "Mobility and the Method: From Shakespeare's Treatise on Mab to Descartes' *Treatise on Man*." In *Textures of Renaissance Knowledge*, edited by Philippa Berry and Margaret Tudeau-Clayton, 137–55. Manchester: Manchester University Press, 2003.

Gatti, Hilary. *Giordano Bruno and Renaissance Science*. Ithaca, NY: Cornell University Press, 1999.

——. "The Natural Philosophy of Thomas Harriot." In *Thomas Harriot: An Elizabethan Man of Science*, edited by Robert Fox, 64–92. Aldershot, England: Ashgate, 2000.

Geddes, Louise. *Appropriating Shakespeare: A Cultural History of "Pyramus and Thisbe."* Madison, NJ: Fairleigh Dickinson University Press, 2017.

Gent, Lucy. "The Self-Cozening Eye." *Review of English Studies* 34, no. 136 (1983): 419–28.

Gillespie, Stuart. *Shakespeare's Books: A Dictionary of Shakespeare Sources*. New Brunswick, NJ: Athlone, 2001.

Goldberg, Benjamin. *The Mirror and Man*. Charlottesville: University Press of Virginia, 1985.

Goldberg, Jonathan. "Dover Cliff and the Conditions of Representation: *King Lear* 4:6 in Perspective." *Poetics Today* 5, no. 3 (1984): 537–47.

Gowland, Angus. *The Worlds of Renaissance Melancholy: Robert Burton in Context*. Cambridge: Cambridge University Press, 2006.

Grabes, Herbert. *The Mutable Glass: Mirror-Imagery in Titles and Texts of the Middle Ages and English Renaissance*. Cambridge: Cambridge University Press, 1982.

Grafton, Anthony. *Defenders of the Text: The Traditions of Scholarship in an Age of Science, 1450–1800*. Cambridge, MA: Harvard University Press, 1991.

Granville-Barker, Harley. *Prefaces to Shakespeare*. 4 vols. London: Sidgwick and Jackson, 1927.

Greenblatt, Stephen. "Shakespeare Bewitched." In *New Historical Literary Study: Essays on Reproducing Texts, Representing History*, edited by Jeffrey N. Cox and Larry J. Reynolds, 108–35. Princeton, NJ: Princeton University Press, 1993.

——. *Shakespearean Negotiations: The Circulation of Social Energy in Renaissance England*. Berkeley: University of California Press, 1988.

——. *The Swerve: How the World Became Modern*. New York: W. W. Norton, 2011.

Gudger, E. W. "The Five Great Naturalists of the Sixteenth Century: Belon, Rondelet, Salviani, Gesner, and Aldrovandi: A Chapter in the History of Ichthyology." *Isis* 22, no. 1 (1934): 21–40.

Guenther, Genevieve. *Magical Imaginations: Instrumental Aesthetics in the English Renaissance*. Toronto: University of Toronto Press, 2012.

Gurr, Andrew. "The Tempest as Theatrical Magic." In *Revisiting "The Tempest": The Capacity to Signify*, edited by Silvia Bigliazzi and Lisanna Calvi, 33–42. New York: Palgrave Macmillan, 2014.

Hahmann, Andree. "Epicurus on Truth and Phantasia." *Ancient Philosophy* 35, no. 1 (2015): 155–82.

Halio, Jay. "The Metaphor of Conception and Elizabethan Theories of the Imagination." *Neophilologus* 50, no. 1 (1966): 454–61.

Hall, A. Rupert. *The Revolution in Science, 1500–1750*. London: Routledge, 1983.

——. *The Scientific Revolution, 1500–1800: The Formation of the Modern Scientific Attitude*. Boston: Beacon, 1966.

Hall, Marie Boas. *The Scientific Renaissance, 1450–1630*. New York: Harper and Row, 1966.

Hannaway, Owen. "Laboratory Design and the Aim of Science: Andreas Libavius versus Tycho Brahe." *Isis* 77, no. 4 (1986): 585–610.

Harcourt, Glen. "Andreas Vesalius and the Anatomy of Antique Sculpture." *Representations* 17 (1987): 28–61.

Harkness, Deborah. *The Jewel House: Elizabethan London and the Scientific Revolution*. New Haven, CT: Yale University Press, 2007.

Harvey, E. Ruth. *The Inward Wits: Psychological Theory in the Middle Ages and the Renaissance*. London: Warburg Institute, 1975.

Haskell, Yasmin Annabel, ed. *Diseases of the Imagination and Imaginary Disease in the Early Modern Period*. Turnhout, Belgium: Brepols, 2011.

Heffernan, Carol Falvo. *The Melancholy Muse: Chaucer, Shakespeare and Early Medicine.* Pittsburgh: Duquesne University Press, 1995.

Heilman, Robert Bechtold. *This Great Stage: Image and Structure in "King Lear."* Baton Rouge: University of Louisiana Press, 1948.

Helgerson, Richard. *The Elizabethan Prodigals.* Berkeley: University of California Press, 1976.

Henninger-Voss, M. "Working Machines and Noble Mechanics: Guidobaldo del Monte and the Translation of Knowledge." *Isis* 91, no. 2 (2000): 233–59.

Hillman, David. *Shakespeare's Entrails: Belief, Scepticism and the Interior of the Body.* New York: Palgrave Macmillan, 2007.

Hillman, David, and Carla Mazzio, eds. *The Body in Parts: Fantasies of Corporeality in Early Modern Europe.* New York: Routledge, 1997.

Hoeniger, David. *Medicine and Shakespeare in the English Renaissance.* Newark: University of Delaware Press, 1992.

Höfele, Andreas. *Stage, Stake, and Scaffold.* Oxford: Oxford University Press, 2011.

Holmes, Joan Ozark. "No 'Vain Fantasy': Shakespeare's Refashioning of Nashe for Dreams and Queen Mab." In *Shakespeare's "Romeo and Juliet,"* edited by Jay L. Halio, 49–82. Newark: University of Delaware Press, 1995.

Huet, Marie-Hélène. *Monstrous Imagination.* Cambridge, MA: Harvard University Press, 1993.

Hunt, Maurice. "Perspectivism in *King Lear* and *Cymbeline.*" *Studies in the Humanities* 14, no. 1 (1987): 18–31.

——. *Shakespeare's Speculative Art.* New York: Palgrave Macmillan, 2011.

Hunter, Lynette. "Cankers in *Romeo and Juliet*: Sixteenth-Century Medicine at a Figural/Literal Cusp." In *Disease, Diagnosis, and Cure on the Early Modern Stage*, edited by Stephanie Moss and Kaara L. Peterson, 171–85. Hants, England: Ashgate, 2004.

Ide, Richard S. "Theatre of the Mind: An Essay on Macbeth." *ELH* 42, no. 3 (1975): 338–61.

Janson, H. W. "The 'Image Made by Chance' in Renaissance Thought." In *De Artibus Opuscula XL: Essays in Honor of Erwin Panofsky*, edited by Millard Meiss, 256–58. New York: New York University Press, 1961.

Johnson, Kimberly. *Made Flesh: Sacrament and Poetics in Post-Reformation England.* Philadelphia: University of Pennsylvania Press, 2014.

Jones, Howard. *The Epicurean Tradition.* London: Routledge, 1989.

Kalas, Rayna. *Frame, Glass, Verse: The Technology of Poetic Invention in the English Renaissance.* Ithaca, NY: Cornell University Press, 2007.

Kaufmann, Thomas DaCosta. *Arcimboldo: Visual Jokes, Natural History, and Still-Life Painting.* Chicago: University of Chicago Press, 2009.

——. *The Mastery of Nature: Aspects of Art, Science, and Humanism in the Renaissance.* Princeton, NJ: Princeton University Press, 1993.

Kemp, Martin. "From 'Mimesis' to 'Fantasia': The Quattrocento Vocabulary of Creation, Inspiration and Genius in the Visual Arts." *Viator* 8 (1977): 347–98.

——. *Leonardo da Vinci: The Marvellous Works of Nature and Man.* Cambridge, MA: Harvard University Press, 1981.

——. *The Science of Art: Optical Themes in Western Art from Brunelleschi to Seurate.* New Haven, CT: Yale University Press, 1990.

Kemp, Simon. *Cognitive Psychology in the Middle Ages.* Westport, CT: Greenwood, 1996.

Kennedy, Judith M., and Richard F. Kennedy, eds. *A Midsummer Night's Dream.* Shakespeare: The Critical Tradition. London: Athlone, 1999.

Kiefer, Frederick. "'Written Troubles of the Brain': Lady Macbeth's Conscience." In *Reading and Writing in Shakespeare*, edited by David M. Bergeron, 64–81. Newark: University of Delaware Press, 1996.

Kinney, Arthur F. *Lies like Truth: Shakespeare, Macbeth, and the Cultural Moment.* Detroit: Wayne State University Press, 2001.

——. *Shakespeare and Cognition: Aristotle's Legacy and Shakespearean Drama.* New York: Routledge, 2006.

Knight, G. Wilson. *The Crown of Life: Essays in Interpretation of Shakespeare's Final Plays.* Oxford: Oxford University Press, 1947.

——. *The Wheel of Fire.* London: Routledge, 2001.

Kuhn, Thomas S. *The Structure of Scientific Revolutions.* Chicago: University of Chicago Press, 2012.

Kusukawa, Sachiko. "Leonhart Fuchs and the Importance of Pictures." *Journal of the History of Ideas* 58, no. 3 (1997): 403–27.

Langley, Eric. "'And Died to Kiss His Shadow': The Narcissistic Gaze in Shakespeare's *Venus and Adonis.*" *Modern Language Studies* 44, no. 1 (2008): 12–26.

Levao, Ronald. "Bacon and the Mobility of Science." *Representations* 40 (1992): 1–32.

Lewis, Cynthia. "'We Know What We Know': Reckoning in *Love's Labor's Lost.*" *Studies in Philology* 105, no. 2 (2008): 245–64.

Lindberg, David C. *Theories of Vision from Al-Kindi to Kepler.* Chicago: University of Chicago Press, 1976.

Lindberg, David C., and Robert S. Westman, eds. *Reappraisals of the Scientific Revolution.* Cambridge: Cambridge University Press, 1990.

Lupton, Julia Reinhard. "Creature Caliban." *Shakespeare Quarterly* 51, no. 1 (2000): 1–23.

MacDonald, Michael. *Mystical Bedlam: Madness, Anxiety, and Healing in Seventeenth-Century England.* Cambridge: Cambridge University Press, 1981.

Mahoney, Charles. "Coleridge and Shakespeare." In *The Oxford Handbook of Samuel Taylor Coleridge*, edited by Frederick Burwick, 498–514. Oxford: Oxford University Press, 2009.

Marchitello, Howard. *The Machine in the Text: Science and Literature in the Age of Shakespeare and Galileo.* Oxford: Oxford University Press, 2011.

Marchitello, Howard, and Evelyn Tribble, eds. *The Palgrave Handbook of Early Modern Literature and Science.* London: Palgrave Macmillan, 2017.

Maurette, Pablo. "De rerum textura: Lucretius, Fracastoro, and the Sense of Touch." *Sixteenth Century Journal* 45, no. 2 (2014): 309–30.

Maus, Katharine Eisaman. *Inwardness and Theater in the English Renaissance.* Chicago: University of Chicago Press, 1995.

——. "Sorcery and Subjectivity in Early Modern Discourses of Witchcraft." In *Historicism, Psychoanalysis, and Early Modern Culture*, edited by Carla Mazzio and Douglas Trevor, 325–48. New York: Routledge, 2000.

Mazzio, Carla. "Shakespeare and Science, c. 1600." *South Central Review* 26 (2009): 1–23.

McCreary, Eugene P. "Bacon's Theory of Imagination Reconsidered." *Huntington Library Quarterly* 36, no. 4 (1973): 317–26.

McElveen, Idris Baker. *Shakespeare and Renaissance Concepts of the Imagination.* Ann Arbor, MI: University Microfilms, 1979.

McGee, John. "Shakespeare's Narcissus: Omnipresent Love in *Venus and Adonis*." *Shakespeare Survey* 63 (2003): 272–81.

McGinnis, Jon. *Avicenna.* Oxford: Oxford University Press, 2010.

Mead, Stephen X. "Shakespeare's Play with Perspective: Sonnet 24, *Hamlet, Lear*." *Studies in Philology* 109, no. 3 (2012): 225–57.

Meek, Richard. *Narrating the Visual in Shakespeare.* Farnham, England: Ashgate, 2009.

Melchior-Bonnet, Sabine. *The Mirror: A History.* Translated by Katharine H. Jewett. New York: Routledge, 2001.

Montrose, Louis. "'Sport by Sport O'erthrown': *Love's Labour's Lost* and the Politics of Play." *Texas Studies in Language and Literature* 18, no. 4 (1976): 528–52.

Moore, Mary B. "Wonder, Imagination, and the Matter of Theatre in *The Tempest*." *Philosophy and Literature* 30, no. 2 (2006): 496–511.

Muir, Kenneth. "Image and Symbol in Macbeth." *Shakespeare Survey* 19 (1966): 45–54.

——. *The Sources of Shakespeare's Plays.* London: Methuen, 1977.

Munro, Lucy. "'Antique/Antic: Archaism, Neologism and the Play of Shakespeare's Words in *Love's Labour's Lost* and *2 Henry IV*." In *Shakespeare's World of Words*, edited by Paul Yachnin, 77–101. London: Bloomsbury, 2015.

Nauta, Lodi, and Detlev Pätzold, eds. *Imagination in the Later Middle Ages and Early Modern Times.* Leuven, Belgium: Peeters, 2004.

Neely, Carol Thomas. *Distracted Subjects: Madness and Gender in Shakespeare and Early Modern Culture.* Ithaca, NY: Cornell University Press, 2004.

Nordlund, Marcus. *The Dark Lantern: A Historical Study of Sight in Shakespeare, Webster, and Middleton.* Göteborg, Sweden: Acta Universitatis Gothoburgensis, 1999.

Nuttall, A. D. *Shakespeare the Thinker.* New Haven, CT: Yale University Press, 2007.

Ogilvie, Brian W. *The Science of Describing: Natural History in Renaissance Europe.* Chicago: University of Chicago Press, 2006.

Ortiz, Joseph M. *Shakespeare and the Culture of Romanticism.* Farnham, England: Ashgate, 2013.

Osler, Margaret J. "Baptizing Epicurean Atomism: Pierre Gassendi on the Immortality of the Soul." In *Religion, Science, and Worldview: Essays in Honor of Richard S. Westfall*, edited by Margaret J. Osler and Paul L. Farber, 163–83. Cambridge: Cambridge University Press, 1985.

——, ed. *Rethinking the Scientific Revolution.* Cambridge University Press, 2000.

O'Sullivan, Mary. "Hamlet and Dr. Timothy Bright." *PMLA* 41, no. 3 (1926): 667–79.

Ovitt, George, Jr. *The Restoration of Perfection: Labor and Technology in Medieval Culture.* New Brunswick, NJ: Rutgers University Press, 1987.

Pagden, Anthony. *European Encounters with the New World: From Renaissance to Romanticism.* New Haven, CT: Yale University Press, 1993.

Palfrey, Simon. *Poor Tom: Living "King Lear."* Chicago: University of Chicago Press, 2014.

Panofsky, Erwin. *Perspective as Symbolic Form.* Cambridge, MA: MIT Press, 1991.

Park, Katharine, and Lorraine Daston, eds. *Cambridge History of Science.* Vol. 3, *Early Modern Science.* Cambridge: Cambridge University Press, 2006.

Parker, Patricia. "Rude Mechanicals." In *Subject and Object in Renaissance Culture,* edited by Margreta de Grazia, Maureen Quilligan, and Peter Stallybrass, 43–82. Cambridge: Cambridge University Press, 1996.

———. *Shakespeare from the Margins: Language, Culture, Context.* Chicago: University of Chicago Press, 1996.

Pask, Kevin. *The Fairy Way of Writing: Shakespeare to Tolkien.* Baltimore: Johns Hopkins University Press, 2013.

Pasnau, Robert. *Theories of Cognition in the Later Middle Ages.* Cambridge: Cambridge University Press, 1997.

Passannante, Gerard. *The Lucretian Renaissance: Philology and the Afterlife of Tradition.* Chicago: University of Chicago Press, 2011.

Paster, Gail Kern. *The Body Embarrassed: Drama and the Disciplines of Shame in Early Modern England.* Ithaca, NY: Cornell University Press, 1993.

———. *Humoring the Body: Emotions and the Shakespearean Stage.* Chicago: University of Chicago Press, 2004.

Peltonen, Markku, ed. *The Cambridge Companion to Bacon.* Cambridge: Cambridge University Press, 1996.

Pettigrew, Todd H. J. *Shakespeare and the Practice of Physic: Medical Narratives on the Early Modern English Stage.* Newark: University of Delaware Press, 2007.

Phillips, James E. "*The Tempest* and the Renaissance Idea of Man." *Shakespeare Quarterly* 15, no. 2 (1964): 147–59.

Pine, Martin L. *Pietro Pomponazzi: Radical Philosopher of the Renaissance.* Padua, Italy: Editrice Antenore, 1986.

Pizzorno, Patrizia Grimaldi. *The Ways of Paradox from Lando to Donne.* Florence: Leo S. Olschki, 2007.

Platt, Peter G. *Shakespeare and the Culture of Paradox.* Farnham, England: Ashgate, 2009.

Pollard, Tanya. "'A Thing like Death': Potions and Poisons in *Romeo and Juliet* and *Antony and Cleopatra.*" *Renaissance Drama* 32 (2003): 95–121.

Porter, Joseph A. *Shakespeare's Mercutio: His History and Drama.* Chapel Hill: University of North Carolina Press, 1988.

Porter, Roy, ed. *Cambridge History of Science.* Vol. 4, *Eighteenth-Century Science.* Cambridge: Cambridge University Press, 2008.

Purkiss, Diane. *At the Bottom of the Garden: A Dark History of Fairies, Hobgoblins, and Other Troublesome Things.* New York: New York University Press, 2001.

Pye, Christopher. *The Regal Phantasm: Shakespeare and the Politics of Spectacle.* London: Routledge, 1990.

Read, Sophie. *Eucharist and the Poetic Imagination in Early Modern England.* Cambridge: Cambridge University Press, 2013.

Ritchie, Fiona, and Peter Sabor, eds. *Shakespeare in the Eighteenth Century.* Cambridge: Cambridge University Press, 2012.

Rossky, William. "Imagination in the English Renaissance: Psychology and Poetic." *Studies in the Renaissance* 5 (1958): 49–73.

Rzepka, Adam. "'How Easy Is a Bush Supposed a Bear?': Differentiating Imaginative Production in *A Midsummer Night's Dream*." *Shakespeare Quarterly* 66, no. 3 (2015): 308–28.

Sawday, Jonathan. *The Body Emblazoned: Dissection and the Human Body in Renaissance Culture*. London: Routledge, 1995.

Schalkwyk, David. "The Role of Imagination in *A Midsummer Night's Dream*." *Theoria* 66 (1986): 51–65.

Schleiner, Winfried. *Medical Ethics in the Renaissance*. Washington, DC: Georgetown University Press, 1995.

——. *Melancholy, Genius, and Utopia in the Renaissance*. Wiesbaden, Germany: Harrassowitz, 1991.

Schlutz, Alexander M. *Mind's World: Imagination and Subjectivity from Descartes to Romanticism*. Seattle: University of Washington Press, 2009.

Schmitt, C. B., Quentin Skinner, Eckhard Kessler, and Jill Kraye, eds. *Cambridge History of Renaissance Philosophy*. Cambridge: Cambridge University Press, 1988.

Schoenfeldt, Michael C. *Bodies and Selves in Early Modern England: Physiology and Inwardness in Spenser, Shakespeare, Herbert, and Milton*. Cambridge: Cambridge University Press, 1999.

Segal, Charles. *Lucretius on Death and Anxiety: Poetry and Philosophy in "De Rerum Natura."* Princeton, NJ: Princeton University Press, 1990.

Shapin, Steven. "The House of Experiment in Seventeenth-Century England." *Isis* 79, no. 3 (1988): 373–404.

——. "'A Scholar and a Gentleman': The Problematic Identity of the Scientific Practitioner in Early Modern England." *History of Science* 29 (1991): 279–327.

——. *The Scientific Revolution*. Chicago: University of Chicago Press, 1998.

Shickman, Allan R. "The Fool's Mirror in *King Lear*." *English Literary Renaissance* 21, no. 1 (1991): 75–86.

Shuger, Debora. "The 'I' of the Beholder: Renaissance Mirrors and the Reflexive Mind." In *Renaissance Culture and the Everyday*, edited by Patricia Fumerton and Simon Hunt, 21–41. Philadelphia: University of Pennsylvania Press, 1999.

Siegel, Rudolph E. *Galen on Psychology, Psychopathology, and Function and Diseases of the Nervous System*. Basel, Switzerland: Karger, 1973.

——. *Galen on Sense Perception*. Basel, Switzerland: Karger, 1970.

Siraisi, Nancy G. *The Clock and the Mirror: Girolamo Cardano and Renaissance Medicine*. Princeton, NJ: Princeton University Press, 1997.

——. *Medieval and Early Renaissance Medicine: An Introduction to Knowledge and Practice*. Chicago: University of Chicago Press, 1990.

Slater, Ann Pasternak. "Macbeth and the Terrors of the Night." *Essays in Criticism* 28, no. 2 (1978): 112–28.

Slights, William E. *The Heart in the Age of Shakespeare*. Cambridge: Cambridge University Press, 2008.

Smidt, Kristian. "Shakespeare in Two Minds: Unconformities in *Love's Labour's Lost*." *English Studies* 65, no. 3 (1984): 205–19.

Smith, A. Mark. *From Sight to Light: The Passage from Ancient to Modern Optics.* Chicago: University of Chicago Press, 2015.

Smith, Bruce R. "Speaking What We Feel about *King Lear.*" In *Shakespeare, Memory, and Performance,* edited by Peter Holland, 23–42. Cambridge: Cambridge University Press, 2006.

Soellner, Rolf. *Shakespeare's Patterns of Self-Knowledge.* Columbus: Ohio University Press, 1972.

Spellberg, Matthew. "Feeling Dreams in *Romeo and Juliet.*" *English Literary Renaissance* 43, no. 1 (2013): 62–85.

Spiller, Elizabeth. *Science, Reading, and Renaissance Literature: The Art of Making Knowledge, 1580–1670.* Cambridge: Cambridge University Press, 2004.

——. "Shakespeare and the Making of Early Modern Science: Resituating Prospero's Art." *South Central Review* 26, no. 1 (2009): 24–41.

Stallybrass, Peter, and Allon White. *The Politics and Poetics of Transgression.* London: Methuen, 1986.

Stavig, Mark. *The Forms of Things Unknown: Renaissance Metaphor in "Romeo and Juliet" and "A Midsummer Night's Dream."* Pittsburgh: Duquesne University Press, 1995.

Stewart, Susan. *Poetry and the Fate of the Senses.* Chicago: University of Chicago Press, 2002.

Stoll, Abraham. "Macbeth's Equivocal Conscience." In *Macbeth: New Critical Essays,* edited by Nick Moschovakis, 132–50. New York: Routledge, 2008.

Streete, Adrian. "Lucretius, Calvin, and Natural Law in *Measure for Measure.*" In *Shakespeare and Early Modern Religion,* edited by David Loewenstein and Michael Witmore, 131–54. Cambridge: Cambridge University Press, 2015.

Swan, Claudia. *Art, Science, and Witchcraft in Early Modern Holland: Jacques de Gheyn II (1565–1629).* Cambridge: Cambridge University Press, 2005.

Swann, Marjorie. *Curiosities and Texts: The Culture of Collecting in Early Modern England.* Philadelphia: University of Pennsylvania Press, 2001.

Targoff, Ramie. *Posthumous Love: Eros and the Afterlife in Renaissance England.* Chicago: University of Chicago Press, 2014.

Thorndike, Lynn. *History of Magic and Experimental Science.* 8 vols. New York: Columbia University Press, 1966.

Tilmouth, Christopher. "Shakespeare's Open Consciences." *Renaissance Studies* 23, no. 4 (2009): 501–15.

Traister, Barbara Howard. *Heavenly Necromancers: The Magician in English Renaissance Drama.* Columbia: University of Missouri Press, 1984.

Traversi, Derek. *An Approach to Shakespeare.* London: Sands, 1957.

——. *The Literary Imagination: Studies in Dante, Chaucer, and Shakespeare.* East Brunswick, NJ: Associated University Presses, 1982.

Trevor, Douglas. *The Poetics of Melancholy in Early Modern England.* Cambridge: Cambridge University Press, 2004.

Tribble, Evelyn. *Cognition in the Globe: Attention and Memory in Shakespeare's Theatre.* New York: Palgrave Macmillan, 2011.

Turner, Henry S. *The English Renaissance Stage: Geometry, Poetics, and the Practical Spatial Arts, 1580–1630.* New York: Oxford University Press, 2006.

———. "Life Science: Rude Mechanicals, Human Mortals, Posthuman Shakespeare." *South Central Review* 26 (2009): 197–217.

Van Dyke, Christina. "An Aristotelian Theory of Divine Illumination: Robert Grosseteste's *Commentary on the Posterior Analytics*." *British Journal for the History of Philosophy* 17, no. 4 (2009): 685–704.

Vaughan, Alden T., and Virginia Mason Vaughan. *Shakespeare's Caliban: A Cultural History*. Cambridge: Cambridge University Press, 1991.

Vaughan, Virginia Mason. "'Something Rich and Strange': Caliban's Theatrical Metamorphoses." *Shakespeare Quarterly* 36, no. 4 (1985): 390–405.

Vendler, Helen. *The Art of Shakespeare's Sonnets*. Cambridge, MA: Harvard University Press, 1997.

Von Koppenfels, Werner. "*Laesa Imaginatio*, or Imagination Infected by Passion in Shakespeare's Love Tragedies." In *German Shakespeare Studies at the Turn of the Twenty-First Century*, edited by Christa Jansohn, 68–83. Newark: University of Delaware Press.

Walker, Daniel Pickering. *Spiritual and Demonic Magic: From Ficino to Campanella*. University Park: Pennsylvania State University Press, 2000.

Wallace, Karl R. *Francis Bacon on the Nature of Man: The Faculties of Man's Soul: Understanding, Reason, Imagination, Memory, Will, and Appetite*. Urbana: University of Illinois Press, 1967.

Watson, Gerard. *Phantasia in Classical Thought*. Galway: Galway University Press, 1988.

Wells, Stanley. *Shakespeare, Sex, & Love*. Oxford: Oxford University Press, 2010.

Whitney, Elspeth. *Paradise Restored: The Mechanical Arts from Antiquity through the Thirteenth Century*. Philadelphia: American Philosophical Society, 1990.

Witmore, Michael. *Culture of Accidents: Unexpected Knowledges in Early Modern England*. Stanford: Stanford University Press, 2002.

INDEX

CPSIA information can be obtained
at www.ICGtesting.com
Printed in the USA
LVHW09*1339040918
589107LV00003B/14/P